...D: Making It
For You

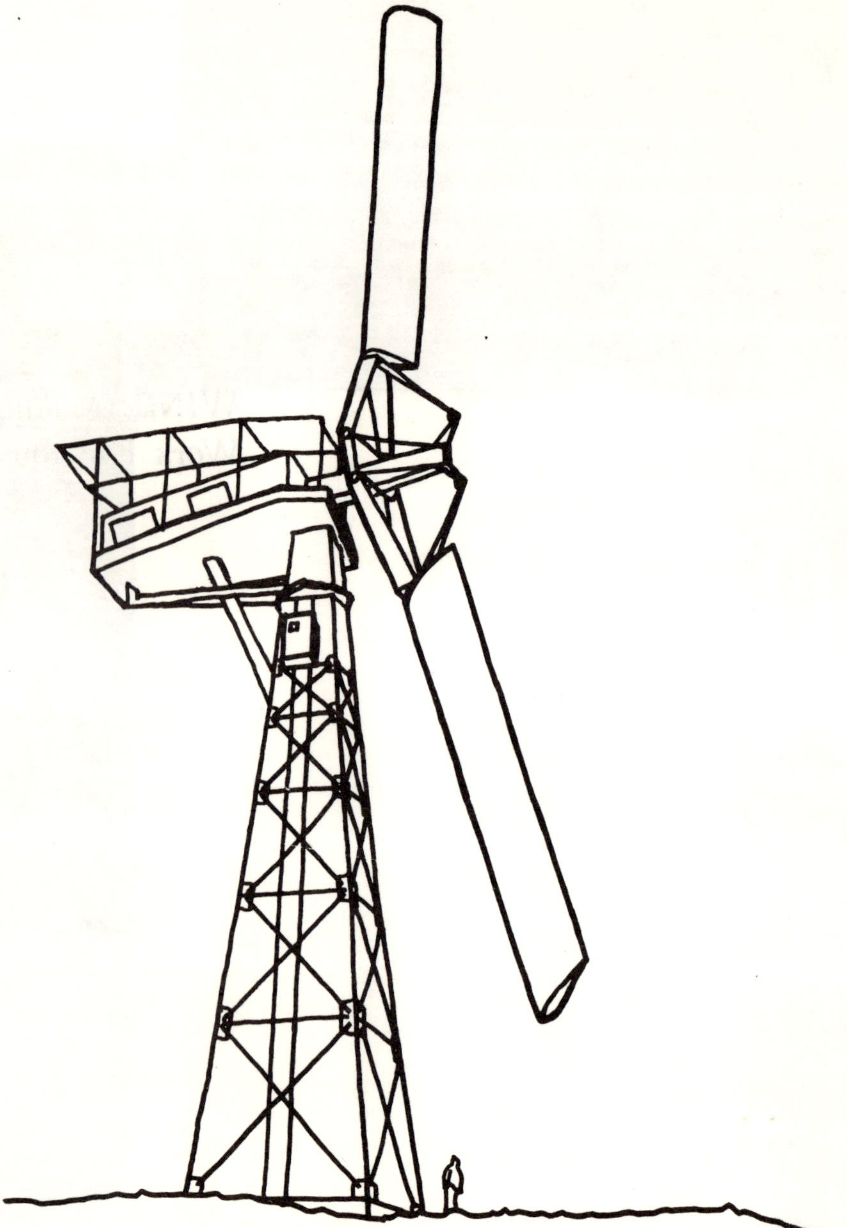

SMITH-PUTNAM TURBINE
GRANPA'S KNOB, VERMONT

The Smith-Putnam wind turbine was the largest wind machine ever built. It was located on Grandpa's Knob near Rutland, Vermont. The collector was 175 feet in diameter (blade tip to blade tip), and the generator was housed on top of a 110 foot tower. This downwind collector generated 1250 kilowatts of power in a 28 mph wind.

Wind

Making It Work For You

Douglas R. Coonley
Cover Art by Don Lackman

THE FRANKLIN INSTITUTE PRESS℠

For Mom and Dad

THE FRANKLIN INSTITUTE PRESS
Philadelphia, Pennsylvania

ISBN Number: 0-89168-008-X
Library of Congress Catalog Card Number: 78-31772
Printed in the United States of America

Contents

A Dutch four-arm windmill is a fine example of an integrated wind energy system. The mill is used for grinding grain or pumping water, and the building has space for the milling activities and for a storehouse, shop, and living quarters for the miller. These windmills are built almost entirely of wood and are often covered with thatch and colorfully decorated.

I. Introduction

Most of you have experienced the delightfully cool feeling of a strong breeze on a hot summer day. Try to imagine what it would be like not to have cool breezes. No wind at all. What would you miss? After all, the wind seems only a small thing to do without. In fact, at times the wind even seems to be a nuisance.

But it would require many electric fans, and huge amounts of electricity to power those fans, to make even a small fraction of the wind energy produced by the interaction of the sun and the earth. Sailboats would find it difficult to sail. Birds and airplanes would find it more difficult to take off and land. It would be more difficult to ventilate buildings. Pollination and propagation of plants would be more restricted. Towns and cities would be stuffy, smelly, and dirty without the cleansing action of the wind. Waves would all but disappear from our lakes and oceans. Your clothes would not dry so quickly on the clothesline, nor could you fly a kite. No longer would you hear the rush of the wind through the trees.

Finally we would miss an important and useful energy source. Our ancestors understood well just how important a source of energy the wind is, perhaps better than we know it today. (A brief history of wind use appears in Appendix D.) Let us take a closer look at what we can do with the wind. And one day in the near future, we will know much more about wind, its behavior, and how best to work with it.

A design for a wind-cooled and -ventilated home in the Minneapolis area. This is an example of an integrated passive wind system.

II. Wind and Its Behavior

Wind is easy to understand in a general sense, yet it is difficult to predict its behavior. It is necessary to understand what the wind is made up of before we can make a more careful analysis of its ways and its changes from moment to moment or place to place. The variations in the wind are important in understanding how we might use the wind and what we can expect of it.

First, we need a useful definition of wind. We might start by defining the wind in very general terms. Wind may be described as "the natural movement of air about the earth." *Natural* indicates that wind is produced without the aid of man-made devices. The interactions of the earth and the sun produce the winds we feel. *Movement* indicates that wind is traveling about and not standing still. And *air* indicates that wind is something very close to us: the very air we breathe.

What is wind more specifically? There are as many ways to describe the wind as there are to experience it. Wind is many things to many people, and we may all describe the wind a little differently. This is fortunate for we can learn quite a lot from each other's description of what the wind is. Some would say that wind is what cools us off on a hot day; keeps kites flying; helps to pollinate plants; makes waves on lakes and oceans; helps airplanes to take off and land; propels sailboats, sailing ships, and iceboats; moves clouds about in the sky; makes trees sway back and forth. Others offer suggestions that wind howls late at night; hinders walking down a city street; sneaks in to cause drafts around windows and doors; makes smoke rise crookedly upward; makes a pinwheel spin; helps to pump water or to generate electricity; causes storm damage; helps to ventilate spaces we use; cleanses cities and towns of dirty air. You can see there are many, many ways to describe what the wind is!

Variations in Wind

Wind varies greatly in speed and direction depending upon time and location. You might walk 100 feet along an exposed mountain ridge and find that the wind speed doubles and its direction reverses. Imagine the change in wind's behavior as you move from a forest to an open field or from a sheltered hillside to a lakeshore. Wind changes with height above the ground, too. Generally, the higher you go the faster the wind and the more steady its direction. Winds are constantly changing direction, both upwards and downwards as well as from side to side. Their speed and direction can change drastically in a matter of a second or less. Wind can blow in one direction at one side of a structure and in the opposite direction at the other side.

The wind changes from day to night and season to season. Winds generally are stronger during the day than at night. Wind speeds are usually highest during the winter and lowest during the summer. Most areas experience winter winds from one direction; summer winds or breezes from another direction; storm winds from yet another. Shoreline areas experience abrupt shifts in wind direction: The wind comes in from the water in the daytime and goes out over the water at night. Winds often move in circles and at great speeds during storms; they disappear completely during a calm.

You can begin to understand better how the wind works by carefully examining certain other aspects of wind itself. Wind may be classified according to how fast it moves. Lack of wind is a *calm*; slow wind a *breeze*; moderate wind a *wind*; fast wind a *storm*! Wind may also be characterized by the way the air moves. Straight and even-moving air is *smooth flow*. Circular and uneven-moving air is *turbulent flow*. The type of flow depends upon the form of the local terrain that the wind moves through.

Site Effects

The characteristics of your location greatly affect the wind conditions you experience. You would expect a much stronger wind on a mountaintop than in the midst of a dense forest. Similarly, there would be a much stronger breeze in the middle of an open field than behind a shelterbelt of trees. Abrupt changes in shape of the local terrain result in turbulent flow. Gradual changes in shape of the local terrain result in smooth flow. Local terrain features include hills, trees, rock outcroppings, earth mounds, mountains, flagpoles, buildings, and structures of all kinds. Anything that can obstruct the wind is a local terrain feature.

Local terrain features may decrease or increase wind speed depending upon their shape, orientation, and interaction with other features. They can make smooth flow turbulent or turbulent flow smooth. Image wind across an open field. If you place a single evergreen tree in the field, the wind will not change markedly. Yet there will be some change because there is a small sheltered area on the side of the tree away from the wind, the leeward side. The area is small since there is turbulence caused by the wind moving around the sides of the tree. The wind comes back quickly to its original line of flow across the field. If you add a few more trees spaced apart to make a line perpendicular to the wind's

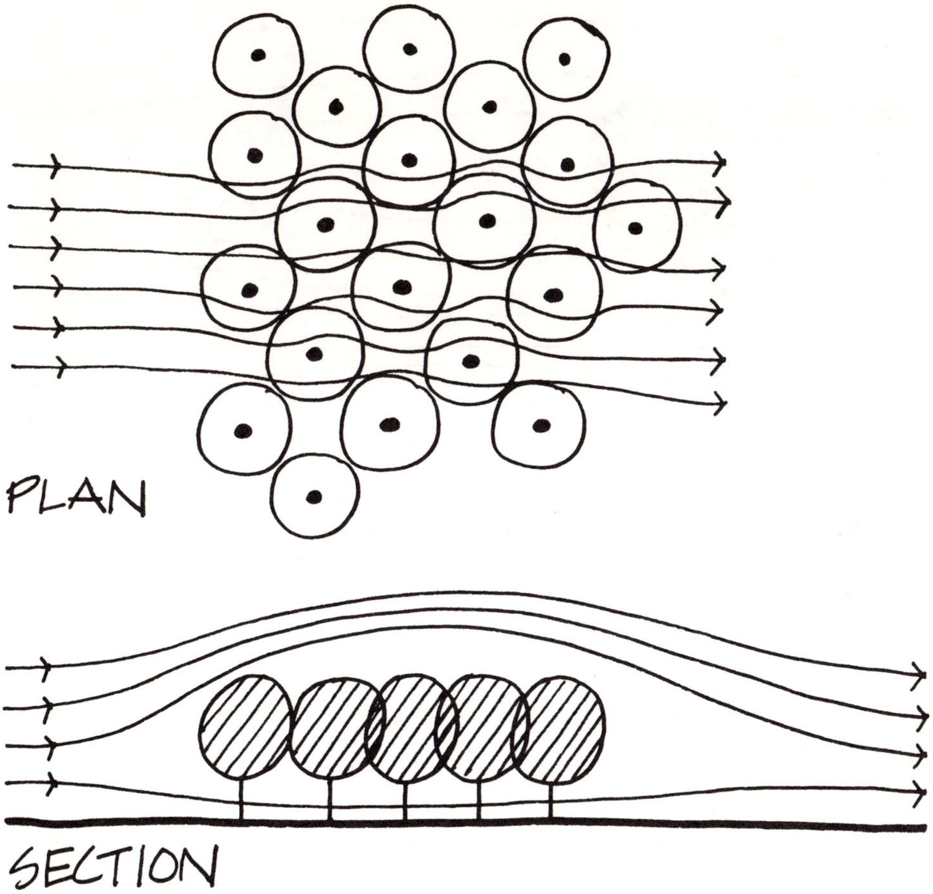

Figure 1. Plan and section views of the airflow through a windbreak of trees.

direction, the wind will flow quickly through the spaces between the trees. There will then be several small sheltered areas (one on the leeward side of each tree) and much turbulence around each tree. The smooth flow of the wind is disturbed by the trees, and more turbulent flow is the result. Now, let's place a few more trees in front of the first row. Let the new trees be in front of the openings between the original trees. Much of the wind must now flow around the new tight clump of trees. The result is a much larger protected area behind the original row of trees. There is much turbulence around the edge of the clump, and some of the wind must go over the top of the clump of trees. Here we have changed smooth flow to turbulent flow and created a large, useful, protected area behind the trees (Figure 1).

Next, imagine filling the same open field with randomly spaced single trees. This changes a very smooth flow into an extremely turbulent flow. If we were to remove several trees in a row through the center of the field, we would have a

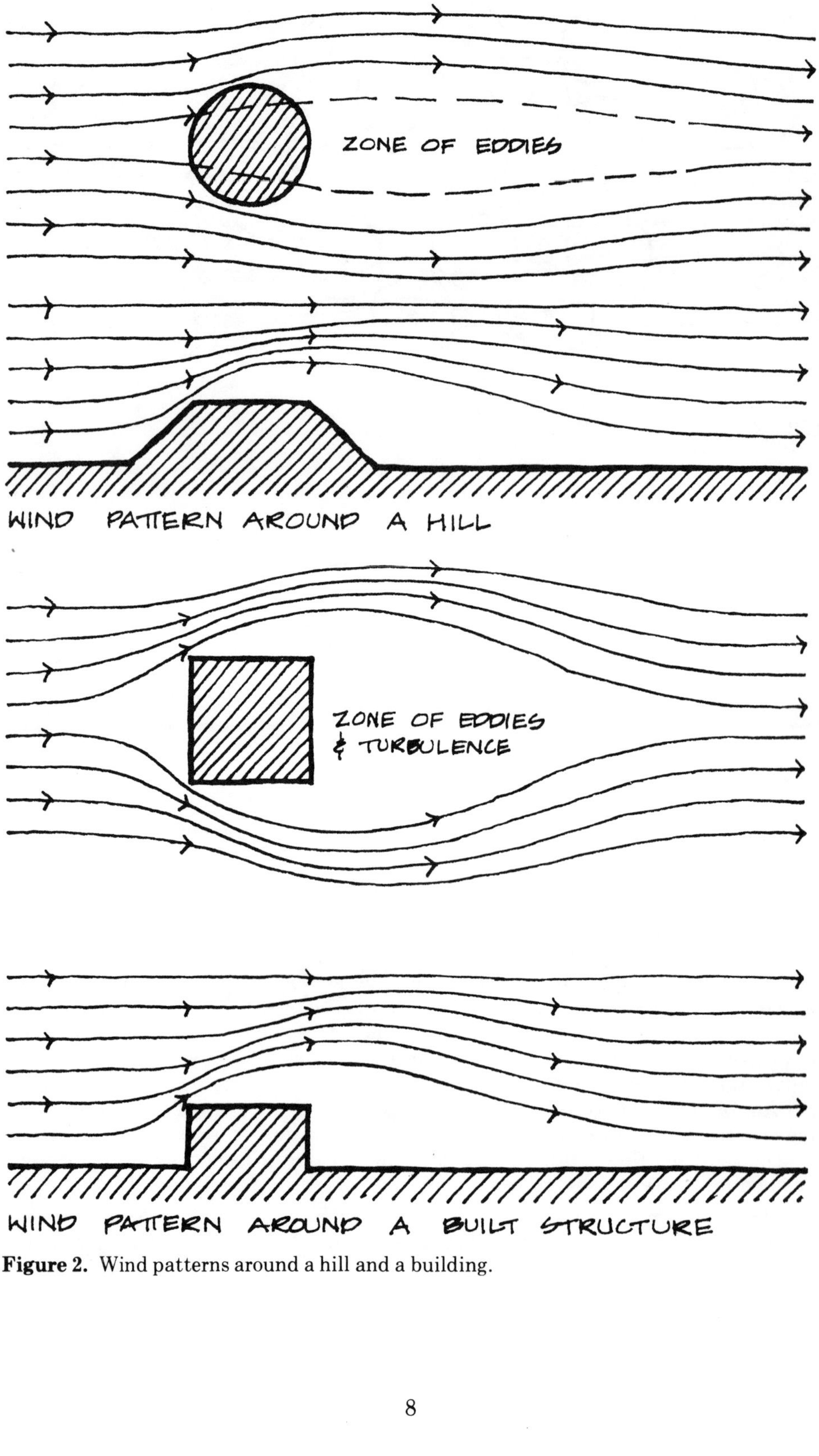

Figure 2. Wind patterns around a hill and a building.

relatively open path for the wind to follow. The wind will flow quite easily along this path through the trees and more smoothly than through the random field of trees. Here the trees change turbulent flow into accelerated smooth flow, an airflow that may be very useful when extracting energy from the wind.

Rocks, hills, mountains, and structures act in ways similar to the trees described above. Rocky areas usually are quite turbulent because of the jagged and irregular shapes of the rocks. Steep mountains and steep or irregular hills generally make turbulence also. Smooth hills or eroded, smooth mountains usually produce smooth, even flow. Sharp corners or edges cause most of the ground-level turbulence in the wind. In order to be reasonably sure of a smooth wind, we must take care not to get too close to a tree line, a cliff edge, or a steep mountain ridge. Man-made structures usually have sharp corners and cause turbulence, but some carefully designed structures such as igloos, tall towers, and old windmills cause very little turbulence or disturbance of airflow. In general, rough terrain disturbs the local airflow near the surface. The disturbance decreases as air moves further away from the surface, be it rough or smooth (Figure 2).

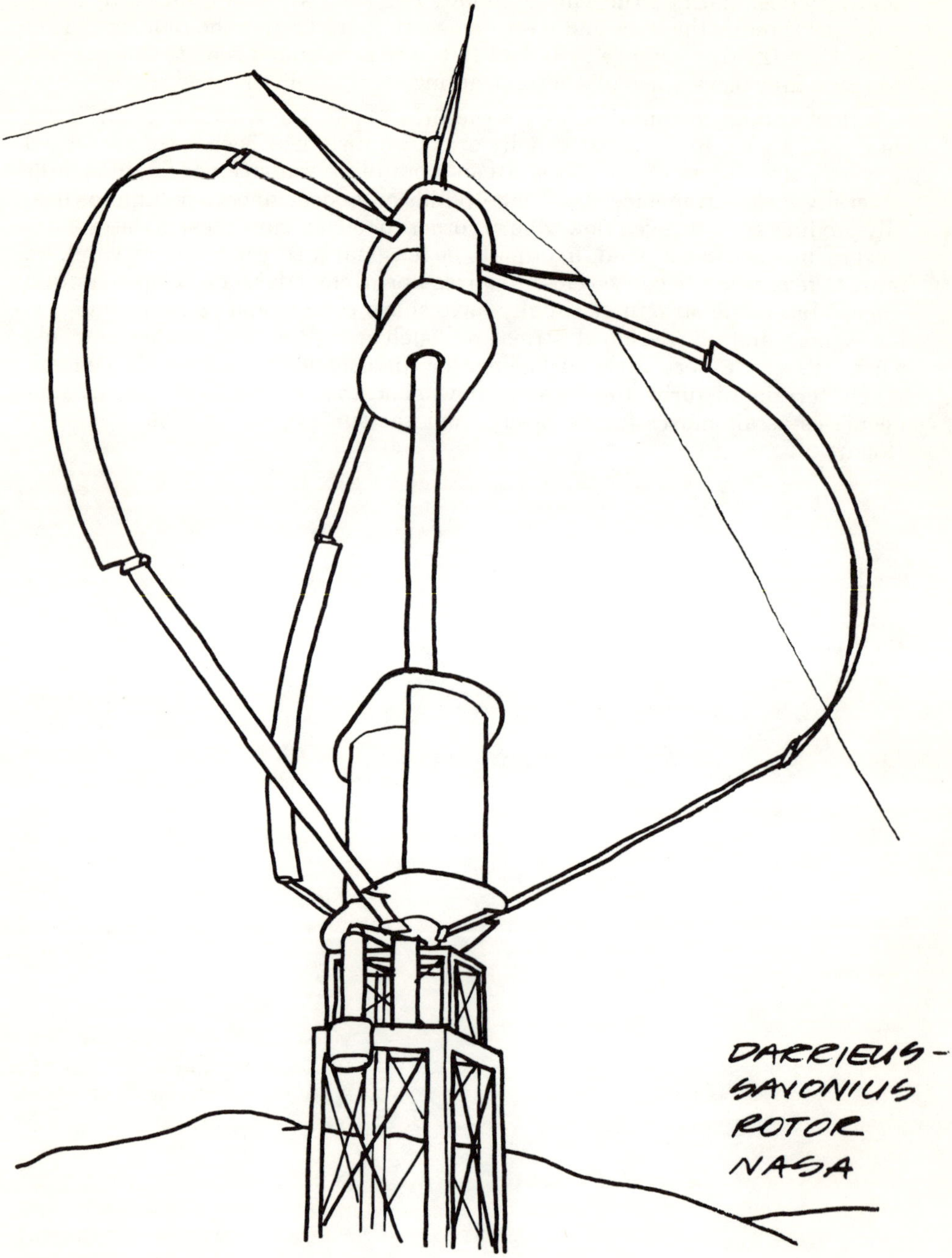

A Darrieus wind collector being developed by Sandia Laboratories and the National Aeronautics and Space Administration (NASA). The Darrieus is a vertical-axis airfoil rotor that is not self-starting. Two Savonius rotors start the Darrieus rotor.

III. How Much Wind Is There?

Measurements of wind speed and direction are important to our understanding and use of wind. It is crucial that we know what sort of wind we have available before we try to work with it, whether it be for ventilation and cooling of our homes or for wind-generated electricity. There are many ways to measure speed and direction of wind. Methods range from continuous monitoring with sophisticated electronic or mechanical equipment to periodic monitoring with less sophisticated equipment. Wind speed is usually measured with a mechanical or electronic device that is powered by the wind and that spins faster or is pushed harder as the wind speed increases. Wind direction is usually measured with a mechanical or electronic device that changes direction with the wind. Wind may be estimated roughly by licking a finger and holding it in the wind or by observing the movement of nearby objects.

For more thorough measurement of wind, we need devices that respond to the wind's changes from moment to moment. Mechanical or electronic anemometers and direction indicators give this kind of response. These devices use continuous recording equipment to give the total wind spectrum. Large weather stations, airports, and highly specialized wind facilities might use this type of wind monitoring.

Measuring the wind and recording data every hour or minute give a wind picture adequate for most purposes. Small weather stations and radio and television stations might use moderate-cost wind-measurement equipment for this type of monitoring (Figure 3). National Weather Service records include both continuous and periodic wind measurement.

Manual recording of wind, whenever it is convenient, gives a more general wind picture. However, these measurements must be carefully coordinated with

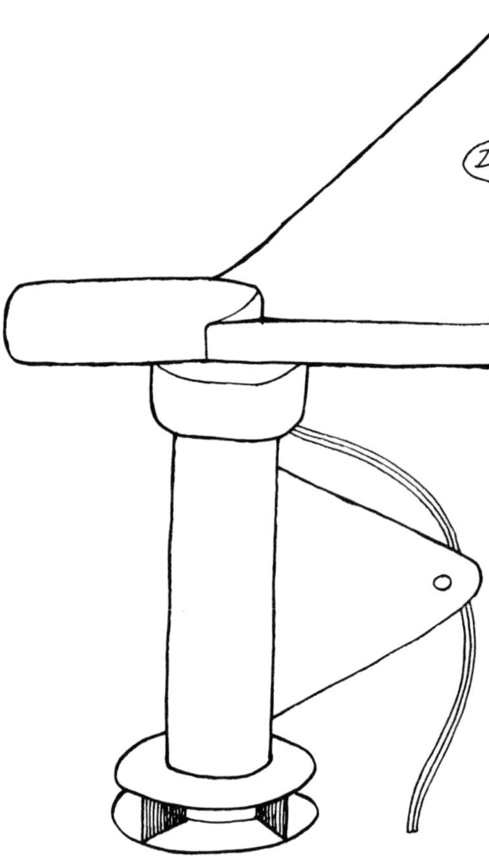

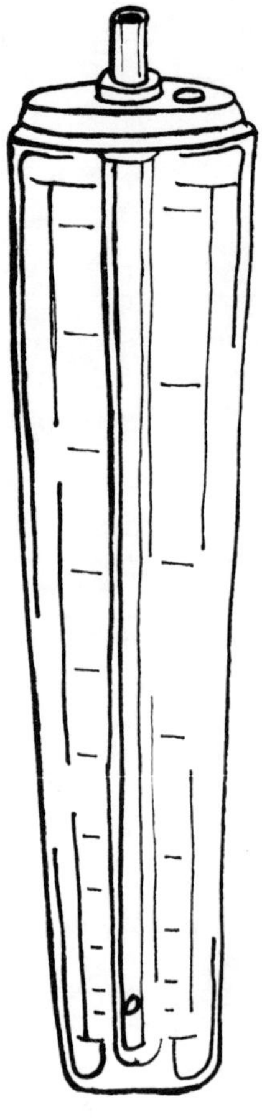

Figure 3. (left) A device designed by Dwyer to measure wind velocity. It is for permanent installation and gives more accurate readings than portable units. The tail vane indicates wind direction.

Figure 4. (right) A hand-held wind-measuring device designed for field use. This unit gives quick readings of wind speed accurate to ten percent, but does not indicate wind direction.

the more accurate wind measurements mentioned above to provide specific information for accurate wind analysis. Do-it-yourselfers and home weather stations, use low-cost measuring devices for this type of wind monitoring. Many different hand-held wind-measuring devices are available in this category; some devices may be carried in a pocket (Figure 4). However, one large problem with this type of equipment is that people are not usually available to read the meters at all hours or during severe weather.

Measurements of wind are not useful in themselves; they must be compiled into a form of wind data. These wind data may then be analyzed and used to estimate the performance of various wind devices we have used or might wish to use. Measurement of wind and compilation of data are recommended for a period of one year for each site analyzed, so we can determine the seasonal variations and have a useful basis for comparison with other places.

Data

Wind data come in many forms ranging from a series of impulses on magnetic tape to colorful charts published by wind enthusiasts. The form the data take usually depends upon the use.

Most wind data are in chart or graph form such as that published by the United States Government in the *Weather Atlas of the United States.* These data are periodically updated and are very useful for preliminary or general design. One major drawback with the weather atlas is its generality and its regional basis. Data are provided only for major cities and are often obtained at airports, which are commonly situated in the windiest areas to facilitate aircraft takeoffs and landings.

More localized wind data may be procured from your local weather station, television or radio station, or airport. These data are usually in a chart or table form and are more localized.

The most accurate data for your site are those that you record and compile yourself or have someone else compile for you. These site data may then be compared with the local and regional data described above. The comparison will provide a total picture of the expected wind conditions for your site over longer periods of time.

Collecting Data

On-site wind data should be gathered on a regular and consistent basis. This will provide the most accurate overall wind picture possible. An automatic recording device will regulate the recording. Don't forget to check the unit periodically to insure that it is in proper working order, has plenty of recording tape, and has an adequate power supply (especially important for battery-powered units). It is a good idea to check the wind-speed and direction indicators regularly, too.

Local wind data may be gathered less regularly from local sources such as a radio or television station or an airport. Check with each source to see when the wind data are available and also how long they are kept on hand. Some radio and television stations keep data for only one day before throwing them away.

General wind data may be gathered at any convenient time from several sources available in most libraries. *The World Almanac,* the *Weather Atlas of the United States* (and other countries), or any natural atlas of the world gives general wind data. The data are commonly given in two forms: chart form and map form. The charts give monthly, yearly, and extreme wind data for certain

Figure 5. Map of the wind energy available in the 48 contiguous states. Sites in the shaded areas have average wind velocities of 10 mph or more at an elevation of 100 feet. Contours are at 2 mph intervals. The bars indicate average wind velocities at specific cities. (Adapted from *Report on Wind Energy*, Federal Power Commission, 1945.)

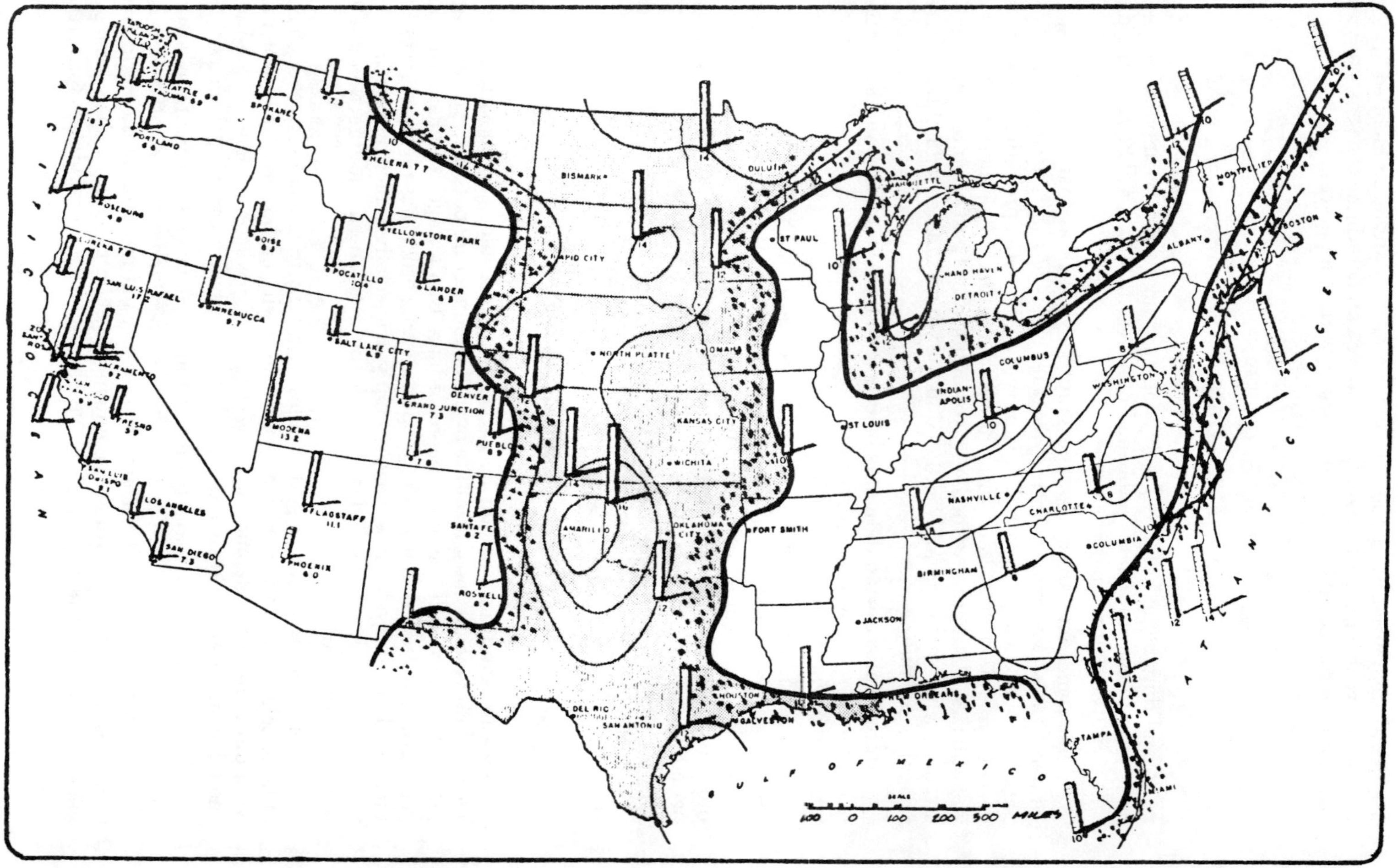

cities. The maps show regional monthly, yearly, and extreme wind data, usually with contours of wind velocity. A wind map of the United States is shown in Figure 5.

Analyzing Data

It is one thing to collect a lot of data, but it is quite another to make the data useful. The wind data must be in an easy-to-use form that gives the information necessary to determine the characteristics of the wind system to be designed.

Wind data may be grouped into short-term data (seconds and minutes), medium-term data (hours and weeks), and long-term data (months and years). Each group of data plays an important part in the design of a wind-utilization system. Short-term data determine the extreme design loads on the system such as wind gust resistance and stability. Medium-term data determine the everyday functioning and wind system sizing. Long-term data determine the total amount of energy expected over a period of time. The wind data must be analyzed carefully in order to realize the required groupings of data.

One quick way to analyze the wind data the first time around is to pick out the extremes. These give a good idea of the maximum wind speeds and directions to be expected. Next, check the day-to-day variations in the wind speed and direction. These determine how large or small the energy storage system should be. Then, find the average wind speed and direction for each month. These averages give an indication of wind variations from season to season, i.e., which months have low wind energy and which have high wind energy. Finally, find the average wind speeds for the entire year and compare it to data for previous years. This indicates the relative amount of energy to be expected from one year to another and over several years' time.

Data analysis depends upon the kind of data collected and their form. Chart or map data are easy to analyze for almost anyone willing to spend a few moments to learn how to read each chart or map. Tables of numbers or graphical wind data may require the use of a calculator and may take longer to analyze. Continuously recorded data on a paper strip chart or magnetic tape require careful analysis by wind specialists; a small computer may be necessary to render the data readable and useful.

Wind-spectrum analyzers are specially designed wind-measuring systems that can be set to stimulate different wind collector systems. These instruments provide useful data about the total energy collected and available for use over whatever period of time the device is monitoring the wind. It usually gives a readout in kilowatt-hours.

Estimating Wind by Observation

Wind, being the movement of air, has a direct effect upon the objects it comes in contact with. By being especially observant of the nearby vegetation,

Table 1. A Useful Scale of Wind Force, Adapted from the Beaufort Scale and Compiled by the US National Weather Service.

Wind Speed (mph)	Wind Effects Observed on Land	Terms Used in NWS Forecasts
0-1	Calm; smoke rises vertically.	Calm
1-3	Direction of wind shown by smoke drift, but not by wind vanes.	Light
4-7	Wind felt on face, leaves rustle, ordinary vane moved by wind.	Light
8-12	Leaves and small twigs in constant motion; wind extends light flag.	Gentle
13-18	Raises dust and loose paper; small branches are moved.	Moderate
19-24	Small trees in leaf begin to sway; crested wavelets form on inland waters.	Fresh
25-31	Large branches in motion; whistling heard in telephone wires; umbrellas used with difficulty.	Strong
32-38	Whole trees in motion; inconvenience felt walking against wind.	Strong
39-46	Breaks twigs off trees; generally impedes progress.	Gale
47-54	Slight structural damage occurs; chimney pots and slates removed.	Gale
55-63	Seldom experienced inland; trees uprooted; considerable structural damage occurs.	Whole gale
64-72	Very rarely experienced; accompanied by widespread damage.	Whole gale
73 or more	Very rarely experienced; accompanied by widespread damage.	Hurricane

we can estimate the wind speed quite accurately. Vegetation is quite responsive to all wind speeds. Trees have light and flimsy leaves, small branches, large limbs, and strong trunks. The leaves respond to the slightest breeze and the trunks only to the strongest gale. Table 1 is a National Weather Service wind-force scale listing some observable effects of the wind.

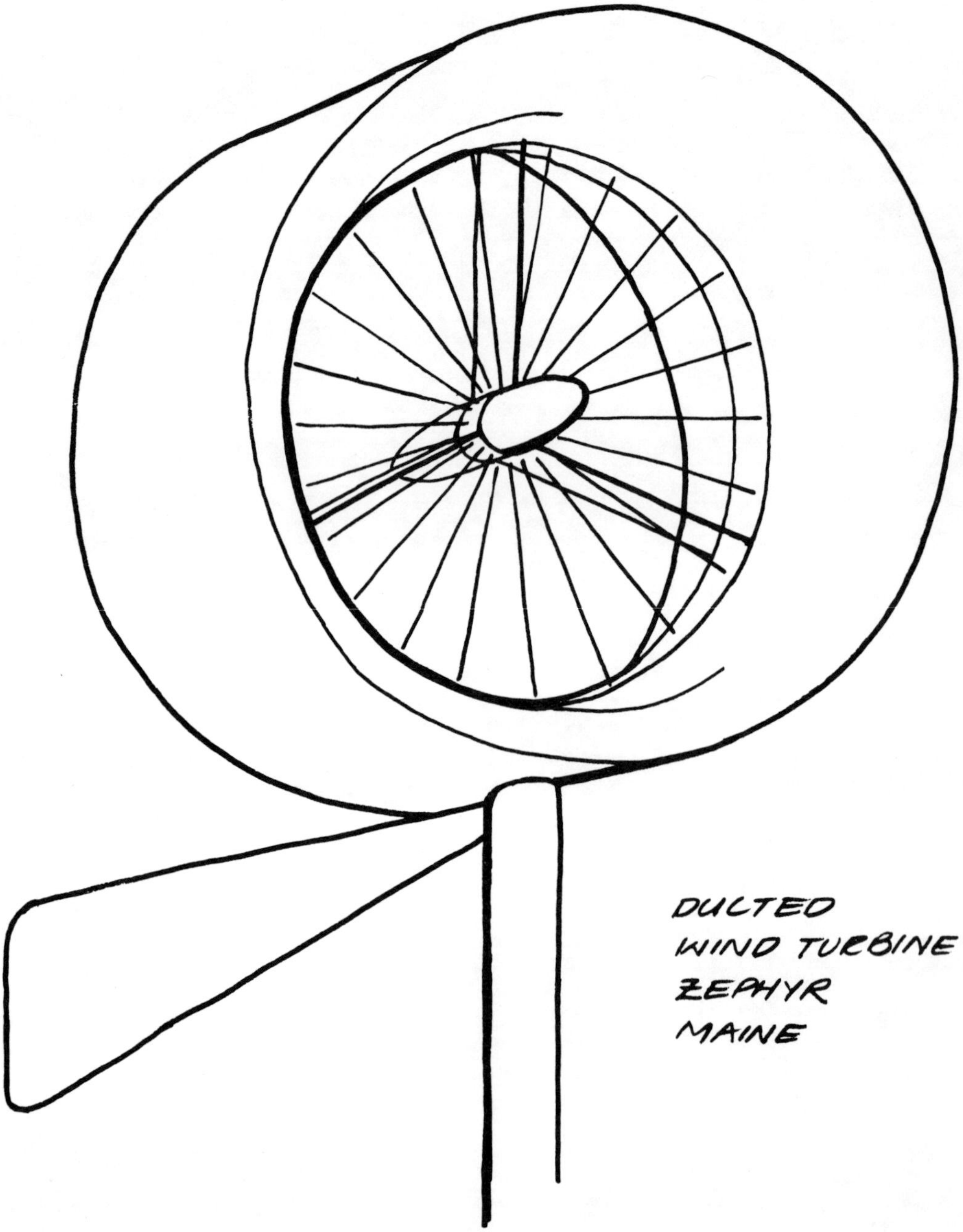

A venturi ducted wind turbine generator developed by the Zephyr Wind Dynamo Company in Maine. This is a multibladed wind collector, and the generator is housed in the duct shroud. The generator has a very large diameter and may rotate slower than a conventional small-diameter generator to produce the same energy output. The tail orients the machine into or out of the wind depending on wind conditions.

IV. Energy in the Wind

Energy available from the wind depends primarily upon three factors: the *velocity* of the wind, the wind collector's *area*, and the *density* of the air that makes up the wind. Wind energy varies differently with each factor, and all are important to the overall performance of a wind system.

Wind energy depends upon the wind *velocity* multiplied by itself twice: This is indicated by $V \times V \times V = V^3$ (velocity cubed). This relationship between wind velocity and wind energy is perhaps the most important relationship to remember when siting and designing a wind system. A simple example will illustrate the importance of this relationship. Wind energy E is proportional to wind velocity cubed V^3. Now, to get an idea of the energy available from the wind at different wind velocities, we can set $E \propto V^3$.* With a wind velocity of five miles per hour we have $E \propto 5 \times 5 \times 5 = 125$. With a wind velocity of ten miles per hour we have $E \propto 10 \times 10 \times 10 = 1000$. There is eight times as much energy available when the wind velocity doubles. In order to receive the most from a wind system, it becomes important to find the speediest wind available. A useful wind velocity graph is shown in Figure 6.

Wind energy depends directly upon the *area* of the wind collector. This area is usually expressed as the "swept area," which varies according to the collectors use and design. The swept area of a wind collector is the largest area swept out by it when it is fully presented to the wind stream. An operating window is a wind collector, and its area is its width times its height. A windmill is a wind collector, and its area is π times its radius squared (πr^2) or the area of the circle swept out by the windmill's rotating blades. A large area collects a larger amount of energy than does a small area. Basically, doubling the area doubles the available energy. This relationship ($E \propto A$) is an important one, too, when designing a wind system.

---

* The symbol $\propto$ means "proportional to."

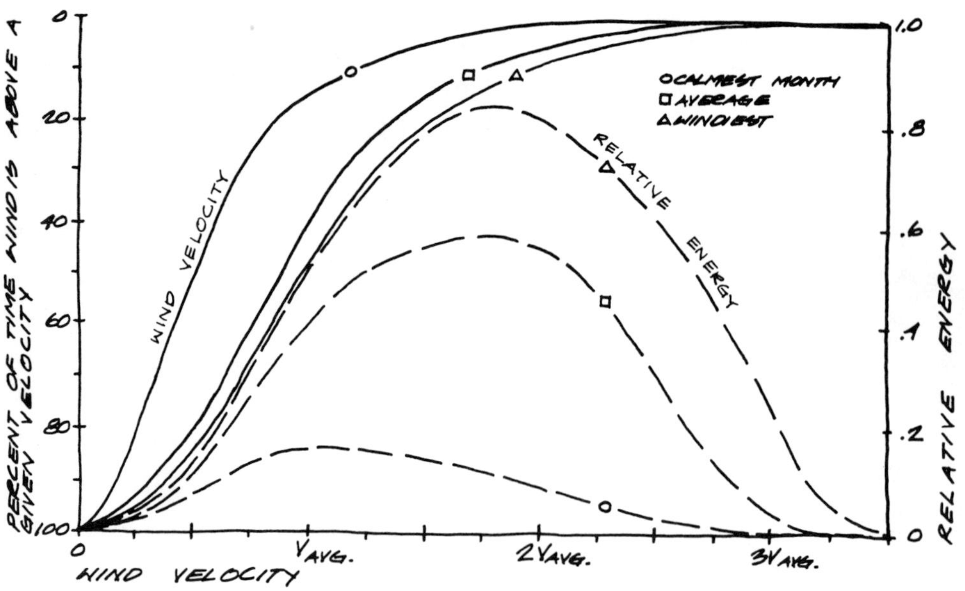

Figure 6. This wind-velocity estimation chart is useful for determining the performance of wind collector systems. The chart shows the wind velocity distribution and relative energy for the calmest, average, and windiest months. (Adapted from Marks', *Standard Handbook for Mechanical Engineers*, 1967.)

To use the chart:

1. Write the average wind velocity for your site at V_{AVG} on the chart.
2. Calculate $2V_{AVG}$ and $3V_{AVG}$ and enter those values on the chart.
3. Determine the minimum velocity at which your collector produces power. The minimum velocity for propeller collectors is usually 7 mph and for drag collectors 5 to 6 mph. Find this value on the horizontal *wind velocity* axis.
4. Draw a vertical line through that point. Find the intersection of that line with the desired wind velocity curve. Usually the curve for the average month is used.
5. Move left from that intersection to the *time* axis.
6. Read the percent of time that the wind collector can be expected to produce energy.
7. Multiply the percentage by the number of hours in the time period you are considering. (One year = 8760 hours.)

See Appendix C for more detailed calculations.

Wind energy also depends upon the *density* of the air that makes up the wind. The more dense the air is, the more energy is available to the wind collector. The air density does not vary substantially from sea level to higher elevations. Yet it does have an effect that should be included in the design of wind systems. This relationship ($E \propto \rho$) is the least important relationship when designing a wind system.

Now, by combining all of these relationships into one, we have:

$$E = V^3 \times A \times \rho \times K.$$

K is a constant that converts units of velocity cubed, area, and density into units of power. And *power* multiplied by *time* equals *energy*. The basic relationship for calculating the power in the wind is given by the equations:

$$P = 1.6 \times 10^3 \times \rho \times A \times V^3$$

for English units: kilowatts, lb/ft^3, ft^2, and ft/sec; and

$$P = 3.2 \times 10^4 \times \rho \times A \times V^3$$

for metric units: kilowatts, kg/m^3, m^2, and m/sec. The conversion constant K changes according to the system of measurement used—English or metric. Be careful to determine which units of measurement are being used. Systems of measurement should not be mixed.

We may now evaluate some different wind situations to determine the amount of energy in the wind. Let us assume that we are located 100 feet above sea level (about the elevation of Philadelphia, Pennsylvania) and that we have a wind collector area of one square foot. We shall assume that our collector can automatically orient itself into the wind as it changes direction. This information allows us to evaluate the energy per unit of time (Wh/h) in the wind for different wind speeds. We can then make a small chart to keep a record of the information as shown in Table 2. If the *average* wind velocity is used to estimate

Table 2. Relationships Among Wind Velocity, Swept Area, Air Density, and Power Available from the Wind.

Location	Philadelphia	Denver	Philadelphia
Elevation (ft)	100	5280	100
Air density (lb/ft^3)	.075	.064	.075
Collector area (ft^2)	1	1	2
Power available in these locations for various wind speeds			
Wind speed (mph)	Power (W)		
0	0	0	0
2.5	1.8	1.5	3.6
5.0	15	13	30
7.5	50	43	100
10.0	120	102	240
12.5	234	199	468
15.0	405	345	810
20.0	960	819	1920
25.0	1875	1600	3750

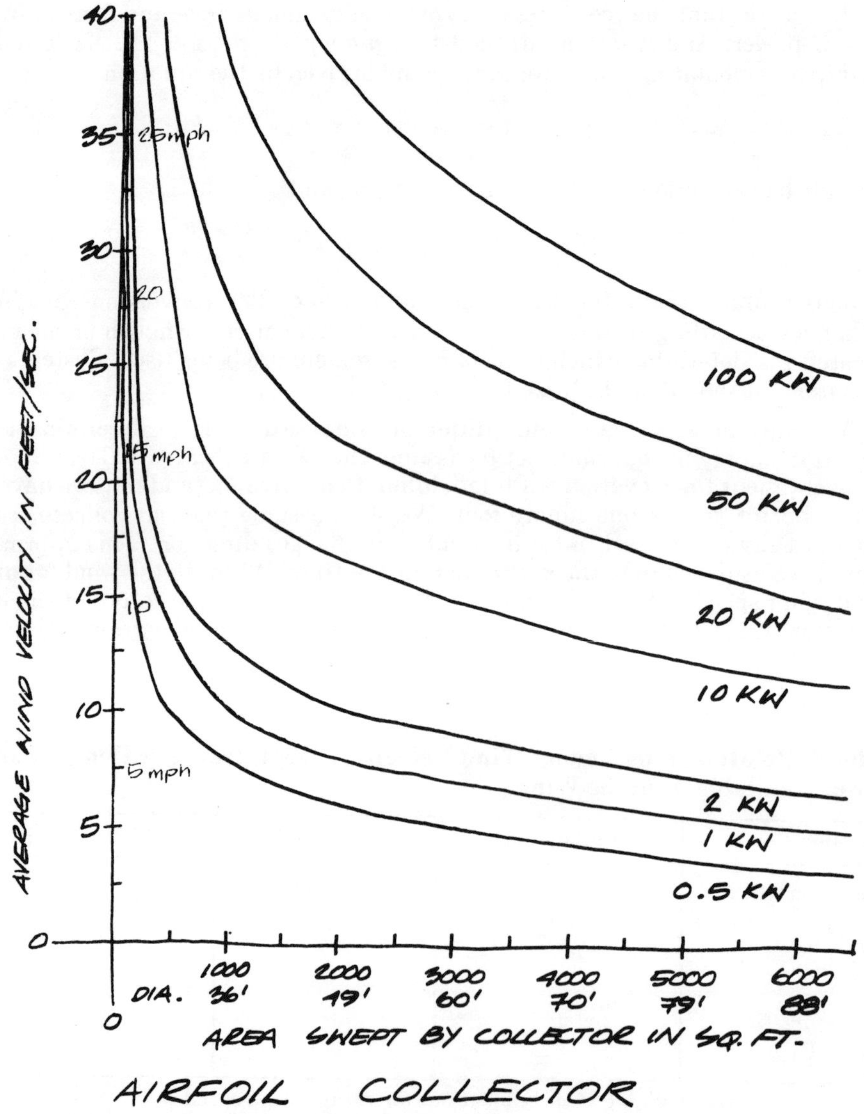

Figure 7. The power output of an airfoil wind collector is shown as a function of average wind velocity and collector area. The *lift* force makes an airfoil collector spin. These collectors have narrow blades that resemble airplane wings. The Smith-Putnam turbine is example of an airfoil collector. (See also Appendix C.)

energy output over a period of time, remember to double the energy amount to account for the increased contribution of higher velocity winds. Available wind power for two basic types of wind collector is shown in Figures 7 and 8.

Table 2 shows that wind energy increases rapidly as the wind velocity increases. It also shows that the energy available varies with the area of the collector. Compare the chart data at 100-foot elevation for collector areas of one

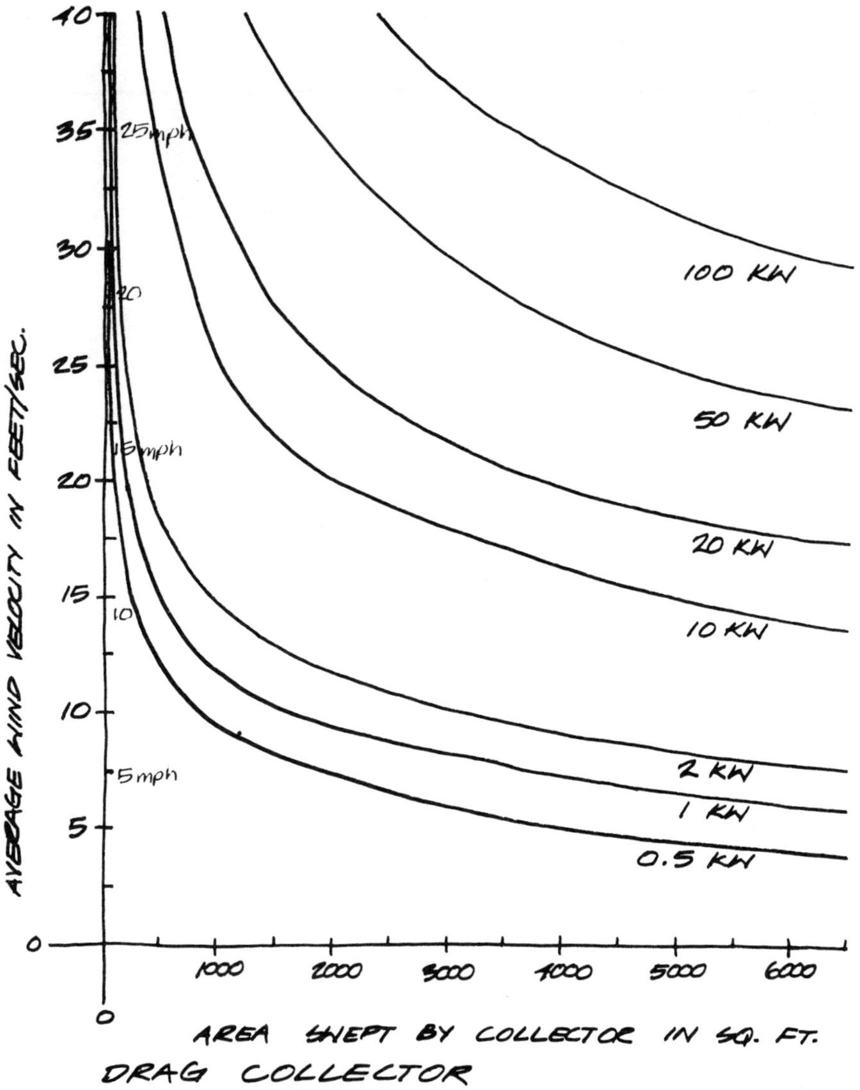

Figure 8. Power output of a drag collector as a function of average wind velocity and collector area. The *drag* force makes these devices spin. A drag collector rotates at the wind speed like an anemometer. The Savonius rotor is another drag collector. (See also Appendix C.)

square foot and two square feet. We can see too how the wind energy varies with elevation above ground by comparing the values for Philadelphia (100 feet) and Denver (5280 feet). We must understand these three factors (wind velocity, collector area, and air density) in designing and using wind energy systems.

Figure 9 is a map showing the energy available from the wind throughout the world. The energy extracted depends on the wind available and the size of

Figure 9. Availability of wind energy for the earth's land areas. The map shows the annual specific energy output of a collector system rated at 25 mph; the output is expressed as kilowatt-hours of energy generated per kilowatt of capacity installed. See text. (Adapted from *Wind Energy Conversion Systems,* NSF-NASA, 1973.)

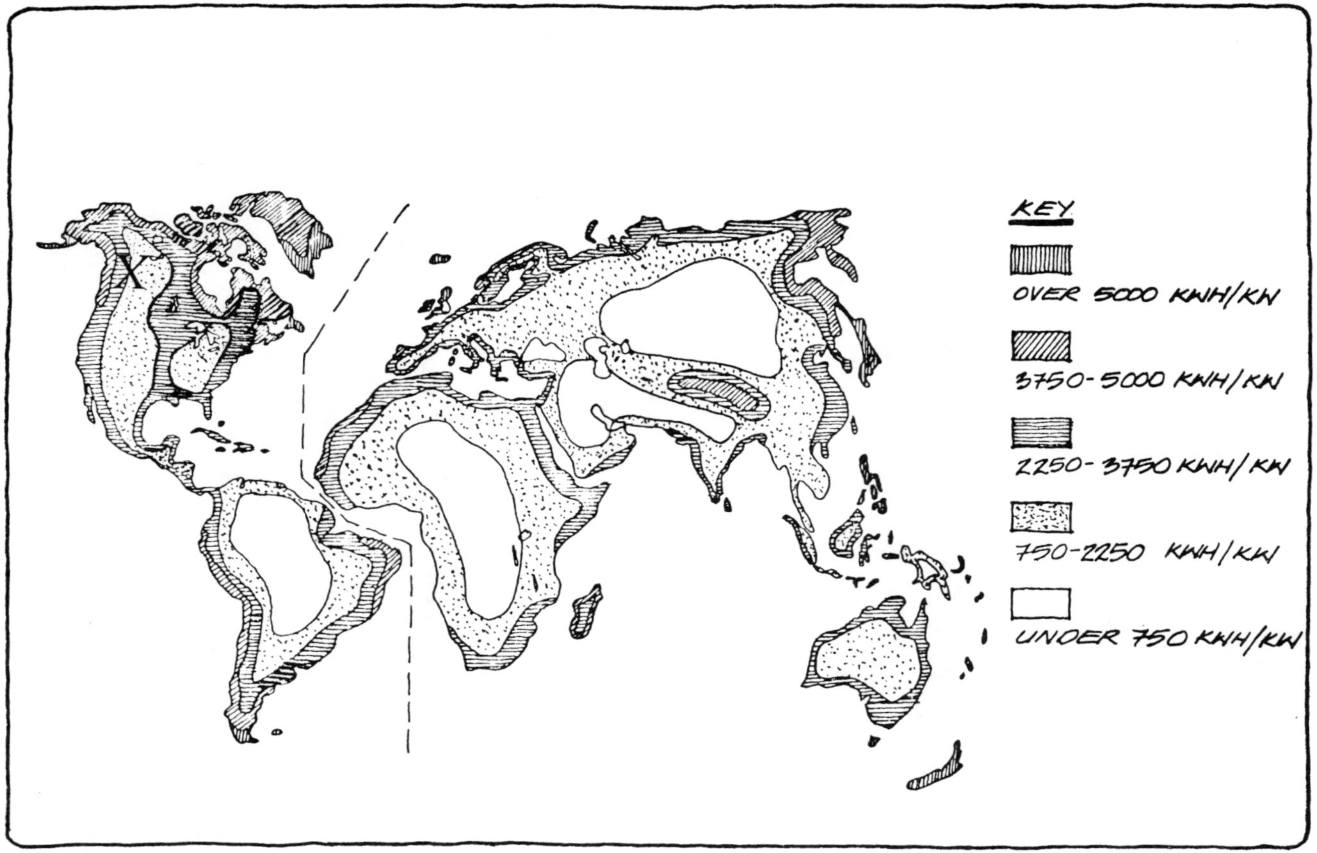

the machine. If the machine is in the Rocky Mountain area (X on the map) and is designed to deliver one kW of power in a 25 mph wind, the average annual energy output will be between 750 kWh and 2250 kWh.

Thus far we have considered only the energy in the wind. The actual usable energy extracted from the wind is much less than the energy in the wind itself (between five percent and 45 percent, depending on the wind system and its use).

A combination water pump and grain mill powered by the wind. One collector performs both jobs simultaneously by means of a mechanical linkage.

V. Using the Wind

Wind may be used for many different purposes depending upon location, available materials, wind speed, and wind direction. Wind systems fall into two general categories: passive systems or active systems.

Passive Systems

Passive systems use wind directly. They usually have very few moving parts and have manual or built-in control systems. Passive systems are susceptible to fluctuations in the wind. Ventilation, cooling, agricultural windmills, ecosystems, transportation (sailboats), and protection are examples of passive uses for wind.

The wind provides ventilation through open windows or doors, air ducts, and vent panels, or even through tiny cracks around the windows and doors. Careful placement of windows and vents can provide adequate ventilation without powered systems. Ventilation is needed year-round, and proper arrangements for preheating the incoming air during the cooler months must be made. A specially designed roof window is shown in Figure 10, and a wind-powered roof ventilator in Figure 11.

Natural cooling of places may be provided by wind through windows, doors, or vent panels. Strategic placement of windows and vents can provide adequate cooling without powered systems even during the warmest months. Cooling may not be needed year-round, and special care should be taken to provide weathertight seals to conserve heat energy during cooler months. A special relationship exists for determining how much cooling is provided by the flow of air. You may also use the bioclimatic chart in Figure 12 to determine how much window and vent area is required to provide adequate cooling, even in the hot-

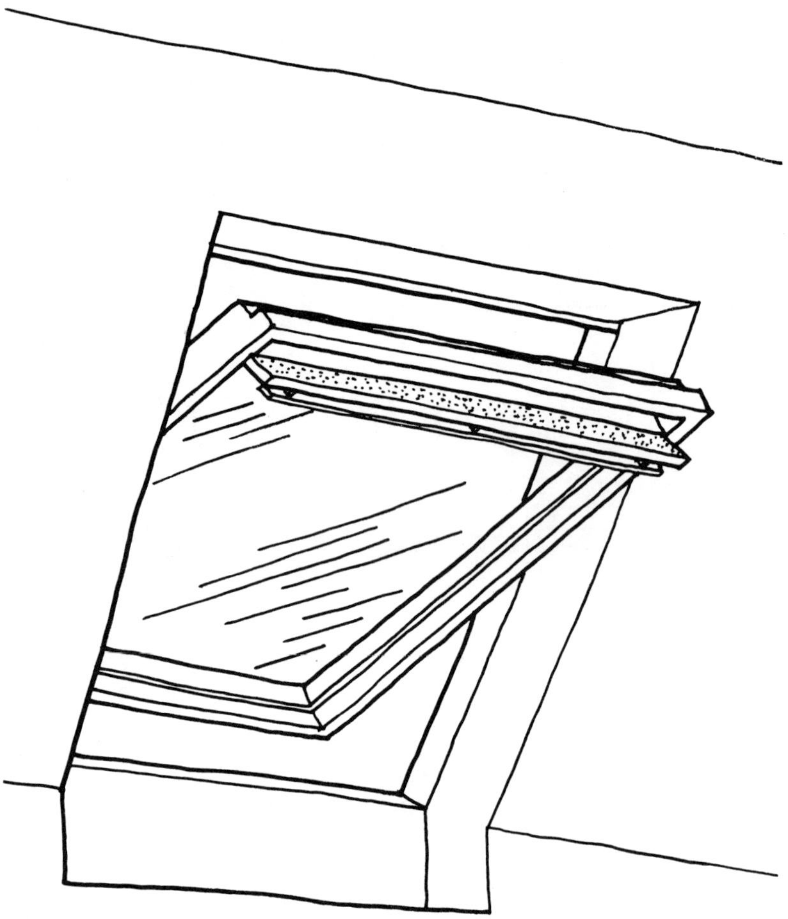

Figure 10. A Velux roof window made in Denmark. The window has a central horizontal pivot to provide natural airflow into and out of the space to be cooled and ventilated. There is a slot at the top of the window that allows ventilation even when the window is closed. This is a useful passive wind system.

test conditions. The placement of windows and vents helps to determine the effectiveness of the natural cooling system, as discussed in Appendix B.

Natural cooling with wind can be successfully coupled with the shading of windows. Windows contribute substantially to heat gain during the warm months, and shades markedly reduce the heat gain. This reduces the need for cooling as well. Usually it is uneconomical to design a wind system to take care of all of the cooling under the most extreme conditions (the hottest day of the year with the highest humidity and little or no wind). Extreme conditions call for basic human ingenuity: remove some clothing or sit in the shade under a water sprinkler.

Agricultural uses for wind are numerous and vary with the character of the area. Winds bring rain for crops and provide means for drying crops after

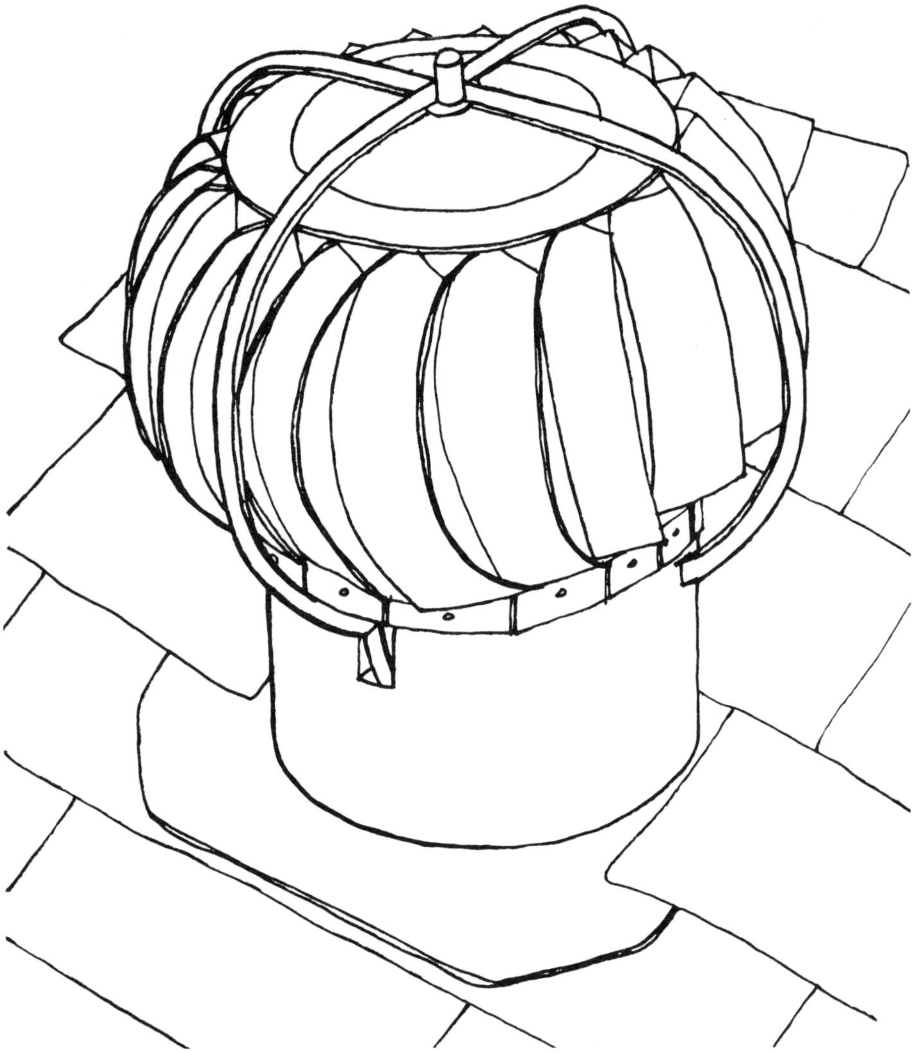

Figure 11. A wind-powered roof ventilator manufactured by Wind Wonder, Inc. This is a vertical-axis wind collector.

harvest; they clean crops by removing dirt and bringing fresh air; they help pollinate plants and crops; they warm or cool growing crops. Strategic placement of fields, based on a knowledge of winds, may inhibit unwanted insects and encourage desirable ones. Winds provide power to activate "passive scarecrows" (noisemakers or tin cans to scare animals or birds away). Smudgepot smoke, insecticides, and fertilizers are distributed evenly through crops by the wind.

Ecosystems rely upon wind for many important activities. Pollination of plants, which was mentioned above, is just one use. Winds help to spread plants by carrying airborne seeds such as maple tree "helicopters" and dande-

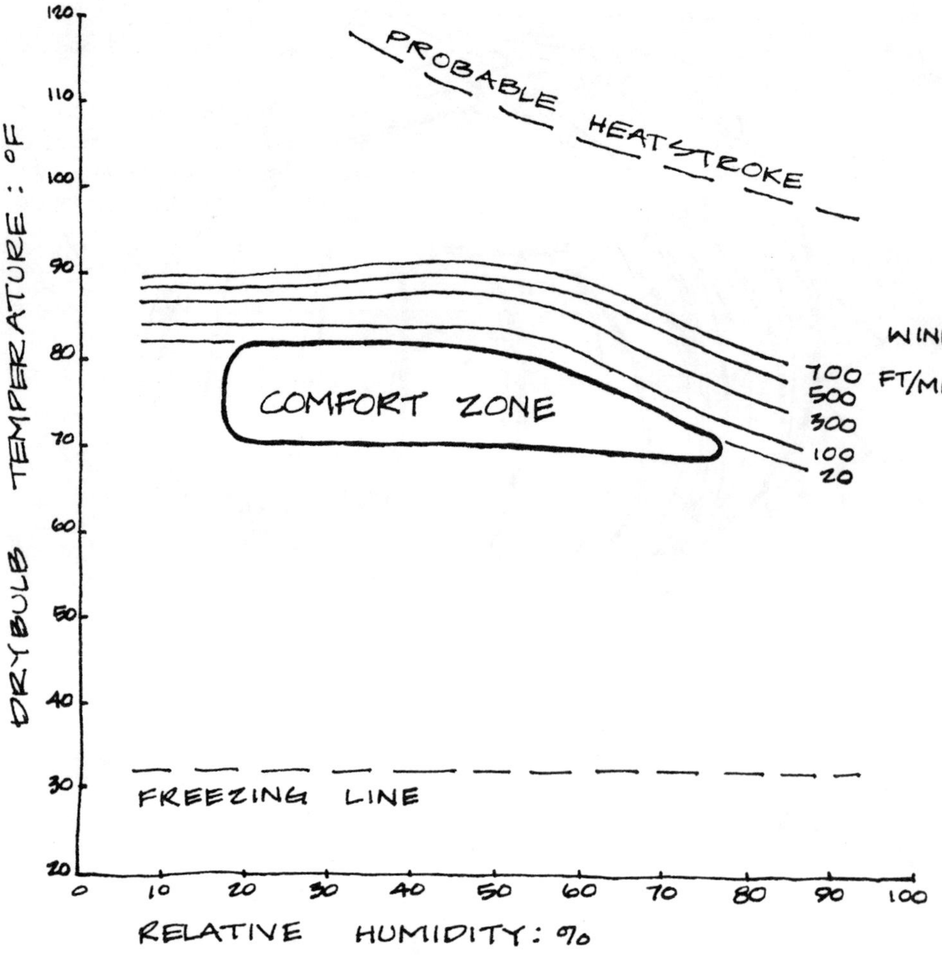

Figure 12. A bioclimatic chart showing how wind cools. Assume that the temperature is 80°F and the relative humidity is 65%. Draw a horizontal line from 80°F on the *temperature* axis and a vertical line from 65% on the *humidity* axis. The intersection of the two lines indicates that wind speed necessary for comfort—in this case 175 ft/min. (Adapted from *Design with Climate* by Victor Olgyay.)

lion "parachutes." Strong winds help to weed out less hardy plants and trees—nature's own method of selective cutting. Storm winds help to remind people where they should and should not build or interfere with nature. Birds and man use the wind to fly. Winds move clouds about and bring rain; they help complete the cycle of water by promoting evaporation from the seas and lakes. Winds contribute to the formation and distribution of soils through erosion and transport of dust. Winds help to produce waves and currents, which provide for oceanic migration of plants and animals; this, in turn, forms the basis for shoreline ecosystems with their constant influxes of fresh water and new life. Another vital wind function is its constant cleansing of the entire surface of the earth and its surrounding atmosphere. The winds are largely responsible for the variations in climate from place to place and for changes in the weather. Arctic and Antarctic winds have a great cooling effect, while tropical winds have a strong warming effect.

Transportation once depended greatly upon winds as a primary source of power. Air travel still depends upon wind for takeoffs and landings, and most airports are located according to the wind characteristics in an area. Most other means of transportation do not depend upon wind directly, and in many cases, wind is regarded as a nuisance since it causes unnecessary drag and impedes progress: Ships are blown off course, airplanes go slower in one direction than another, and automobiles are buffeted around on the highway. However, wind can supply a substantial amount of the driving force required for intercontinental shipping. Projects are now in progress to revive and modernize the clipper ship.

Knowledge of the wind's behavior is also useful in planning for protection. Careful planting of trees and bushes in shelterbelts and windbreaks can reduce excessive wind loads on crops, gardens, yards and buildings. Shelter belts and windbreaks allow passage of some wind at lower velocity and divert most of the wind over and around the windbreak. Evergreens planted as windbreaks against the winter winds can effectively reduce heat losses from wind chill and drafts. In arid regions, earth mounds on the windward sides of fields and buildings can

Figure 13. A Portuguese sail windmill with an inclined axis. This type of windmill is used for grinding grain and pumping water. The sails are furled or unfurled as the wind speed changes. The orientation is set manually; the whole cap of the windmill is turned with a jack.

provide protection from dust storms. The winds are deflected over and around the area protected. Necessary ventilation and cooling are provided while the stronger winds are deflected away.

Active Systems

Active systems use wind indirectly. These systems usually have many moving parts and have automatic and specialized control systems. Active wind systems often have provision for some energy storage so their energy output is less susceptible to fluctuations in the wind than that of passive systems. Milling, water pumping, generation of electricity, and mechanical drive systems are examples of active wind systems.

Figure 14. A Dutch four-arm, hollow post windmill that pumps water for irrigation. This windmill is turned into the wind by a wheel and track mechanism at ground level. The mill is set on a post that rises through the center of the main building; hence the name "hollow post windmill."

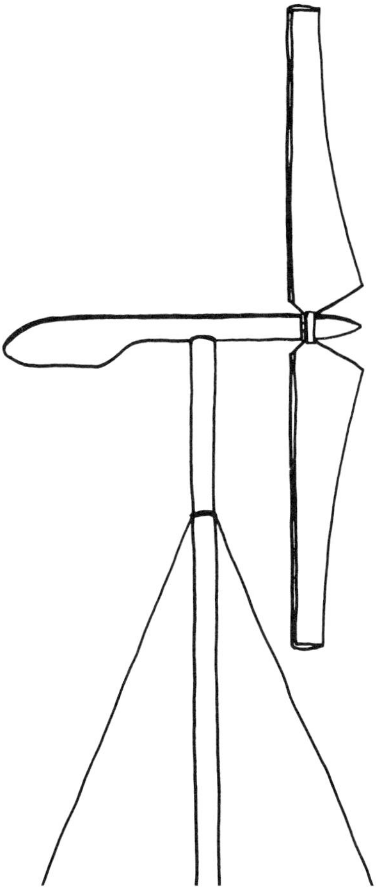

Figure 15. The Princeton Sailwing wind turbine developed by Professor Sweeney. The turbine blades are made of sailcloth stretched over steel spars and pretensioned wires. This is a downwind collector and does not have a tail.

Milling was one of the first active uses for wind, and it is from this use that windmills get their name. Windmills to grind grain were developed centuries ago in Persia. They were built of reeds and wood and powered grinding stones to make meal and flour from grain. Even today windmills are used for this purpose (Figure 13), for sawing wood, and for pumping water. Perhaps the best-known types are the Dutch windmills as illustrated in Figure 14 and in Chapter I.

Wind-powered water pumps have a variety of uses. They provide irrigation, remove water from low-lying areas (such as Holland), and supply domestic or commercial water for farm and household use. Wind-powered water pumps were common in the early days of the United States. The well-known multibladed windmill was used throughout the Midwest and West until the early 1950s when rural electrification programs brought inexpensive and seemingly inexhaustible power to almost everyone. Today many people are refurbishing old water pumps and putting them back to work in order to offset rising energy costs. Wind-powered water pumps like the one illustrated at the beginning of this chapter are still used extensively in remote areas outside of electric utility networks.

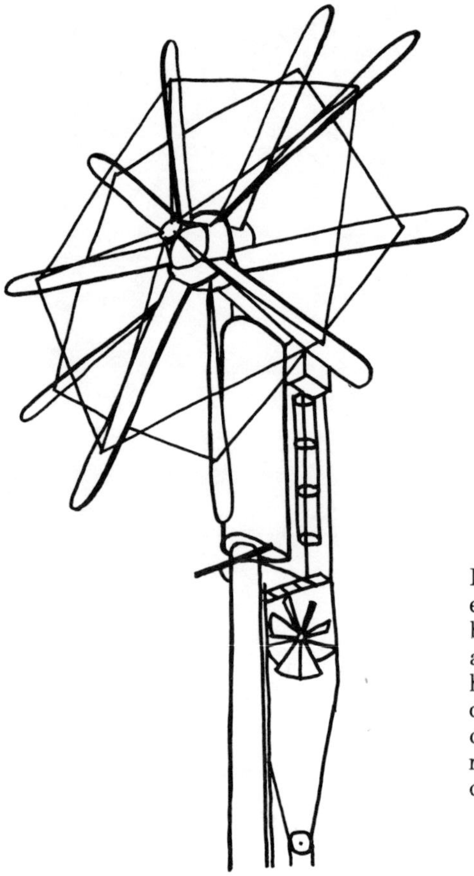

Figure 16. A five-bladed, dual-rotor wind electric system developed by Walter Schönball and located in Germany. Each rotor is attached to one half of a large-diameter hub, and the two rotors spin in opposite directions. The support post has an airfoil cowling to reduce turbulence. The small rotor at the rear automatically orients the collector.

Electrical energy production is one of the most recent uses for wind energy. This is due primarily to the relatively recent development (ca. 1900) and use of electricity as a primary energy source. Some of the first wind-generated electricity came from specially fitted Dutch grinding and water-pumping windmills. Wind electric systems were originally designed to provide electrical power for remote locations where the installation of conventional electric lines was prohibitively expensive and where winds proved to be relatively plentiful. Wind electric systems can generate electricity for specialized uses through solid-state electronic controls. These systems can also generate electricity compatible with conventional powerline electricity through similar electronic equipment (called a "synchronous inverter"). Excess electricity may then be fed directly into the existing electrical power network in a community and thus reduce resource consumption at the main electric power plant. Wind electric systems can also produce direct heat through electric-resistance heating coils or produce gaseous hydrogen through electrolysis of water. Wind electric systems appear to be the most generally compatible form of active wind system due to the overall acceptance and availability of electricity and electric systems. Wind-generated electricity is quickly becoming economically competitive with electricity from

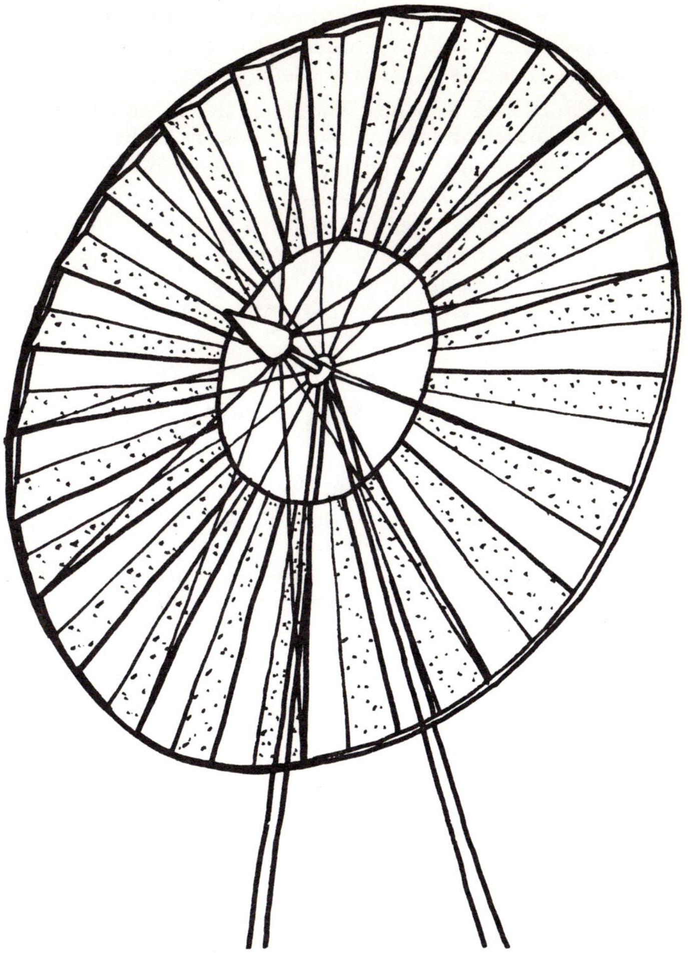

Figure 17. A multiblade, bicycle-wheel turbine developed by Oklahoma State University. Wire spokes stretched between the hub and the rim of the wheel support the rotor blades. The linkage for the generator is located at the rim rather than at the hub.

utility companies. There are many examples of such systems illustrated in this book: They range from the small systems depicted in Figure 15 to the huge Smith-Putnam wind turbine shown in the frontispiece.

Active wind systems can drive various mechanical devices. Almost any piece of stationary machinery can be powered by the wind. Not only grinding and pumping mills, but also wine presses, olive-oil presses, printing presses, sawmills, and hammer mills have used these systems over the years. One mechanical drive system can power several different energy uses at once as in the combination water pump and grinding mill shown at the beginning of this chapter. This system was developed for farm use, but mechanical-drive wind systems can also provide plentiful domestic heat by friction (the mechanical equivalent of heat), which heats a container of fluid such as water or oil. Mechanical wind systems also drive air compressors: The stored compressed air is later used to run air turbines to provide useful energy.

Both passive and active wind systems can be constructed by "do-it-yourselfers." Passive wind systems are usually less complicated and less expensive to build, but they are still subject to variations in the wind. Active wind systems are usually more complicated and expensive to build, yet they react less to variations in the wind because energy storage is usually incorporated into the system design. Both passive and active systems have their places in the overall energy picture. It is important to understand more fully these two types of systems and the implications of each to determine which system is better suited to a given situation.

Available Equipment

Wind systems include windows and doors, small-scale water pumps, and large-scale wind electric systems. Most familiar is the passive wind equipment: operable vent panels, louvers, windows, doors, ducts, roof ventilators, registers, snow fences, sailboats, gliders, and kites. Less familiar is the active wind equipment, which includes wind-powered water pumps, wind collectors of all kinds and sizes, wind electric systems, energy storage systems, and separate wind systems components. This section addresses primarily the group of active wind system components.

Active wind equipment has become much more readily available in the past few years. Perhaps ten manufacturers of wind equipment existed in 1970. Currently, there are more than one hundred manufacturers of wind equipment! The available active wind system equipment ranges in size from two feet in diameter to 35 feet in diameter; in capacity from 10 watts to 25,000 watts (25 kilowatts); and in cost from about $300 to over $25,000.

The choice of active wind systems depends upon three important factors. First, the wind velocity working range of the system determines its applicability to a given site with its specific wind characteristics. Second, the power output of the system determines the amount of energy that can be expected from the wind system over a period of time. Third, the type of collector system determines its suitability for a particular use.

Each active wind collector system performs well within a certain range of wind velocities and conditions. It is especially important to match the wind system with the specific site characteristics in order to make good use of the wind system itself. Active wind systems are designed to give certain energy outputs for specific wind conditions. The overall variation in wind velocities of a specific site determines the amount of energy to be expected over a period of time. Some active wind systems are designed especially for pumping water or compressing air; others are designed to generate electricity or to heat water. The application determines the kind of wind system most suitable for that use. If pumping water is the primary energy need, then a wind-powered water pump is the most suitable system to use. Yet, if both electricity generation and water pumping are the primary uses, a wind electric system that uses some electricity to drive an electric water pump is probably the most suitable system to choose.

There is a wide range of wind equipment available. The systems vary in performance, and the climatic conditions they are designed to perform in also differ. As a result, certain wind systems work better in specific climates. A power output curve for a wind collector system shows the relationship between wind velocity and power output. The power output curve helps in determining the amount of energy to be expected from each wind system, given the characteristics of a certain wind site.

Useful Energy

Useful energy is the total amount of energy provided by a wind system and actually *utilized* to help meet energy needs. Between the wind collector and the actual application of the energy, there are usually many places in the system where energy is lost. Every change in machine speed or in the form of energy (mechanical to electrical; electrical to chemical; chemical to electrical) results

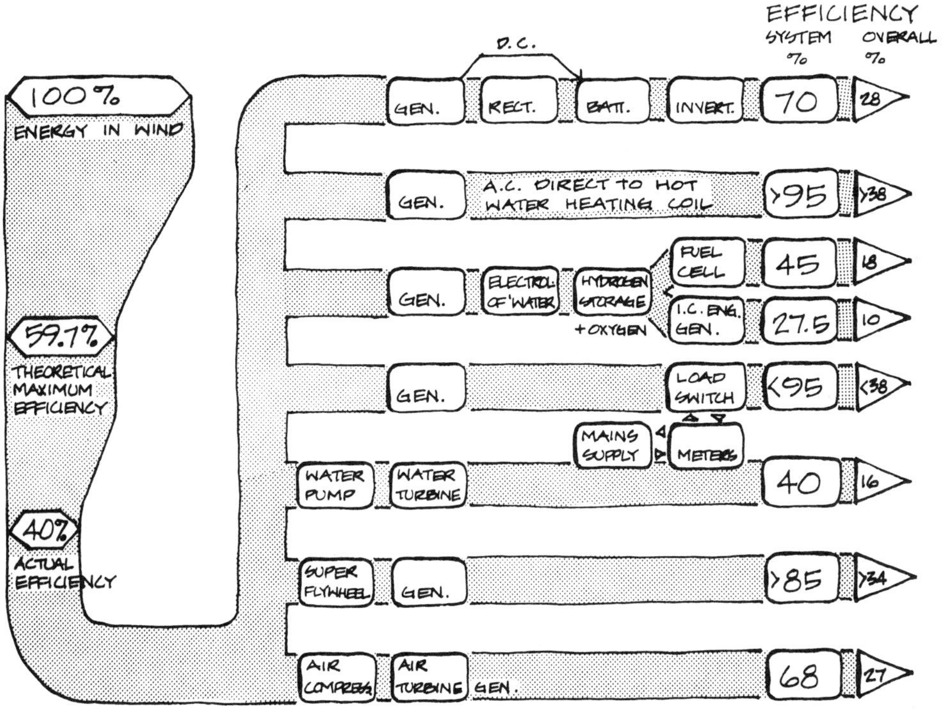

Figure 18. This wind system efficiency chart shows how the various conversion and storage systems cause energy losses. For instance, the actual collector efficiency here is 40 percent. There are eight schemes for converting that energy into usable power. In the first scheme a generator produces DC electricity that is stored in a battery, and an inverter converts the DC current into the more commonly used AC current when needed. This system has an efficiency of 70 percent. Hence the overall efficiency is the product of the collector efficiency and the conversion system efficiency: $.40 \times .70 = .28$, or 28 percent.

in a loss of energy, usually in the form of heat. The overall effect reduces the energy output of the complete wind system. The more direct the system the better is its efficiency from collection to use. For example, a direct-drive wind generator is more efficient than a wind generator that uses a gearbox to increase speed. The fewer stages in the conversion from wind energy to energy use, the smaller the loss along the way. Figure 18 shows roughly the amount of energy that is wasted by components of the wind system and how much one can actually expect to use.

Wind fluctuates between extremes, so there are times when more energy is available than can be used directly or stored for later use. It is necessary to throw away or "dump" some energy at such times; this is another energy loss in the system. Similarly, there are times when there is little or no wind, and the stored energy must be used. There is then a need for an auxiliary system to provide a basic level of energy to keep the system going until the wind system can resume energy production.

The useful energy from a wind system depends on a host of factors, including the number of conversion steps in the system, energy storage, storage capacity, variation of wind velocity and direction, the control-system characteristics, actual energy needs, and variations in energy needs. A wind energy conversion chart for determining the expectable useful energy in the wind is provided in Appendix C. You may find it an interesting project to fill in the chart using data for one of the wind collectors listed in Appendix D. Use wind data for a nearby site. From the chart determine how much auxiliary energy is needed or how much excess energy is available.

Active wind collector systems are made up of many different components that interact to determine the overall system performance. The overall system performance, combined with specific energy needs, determines the useful energy from the wind collector system as shown in the wind energy calculation chart mentioned above.

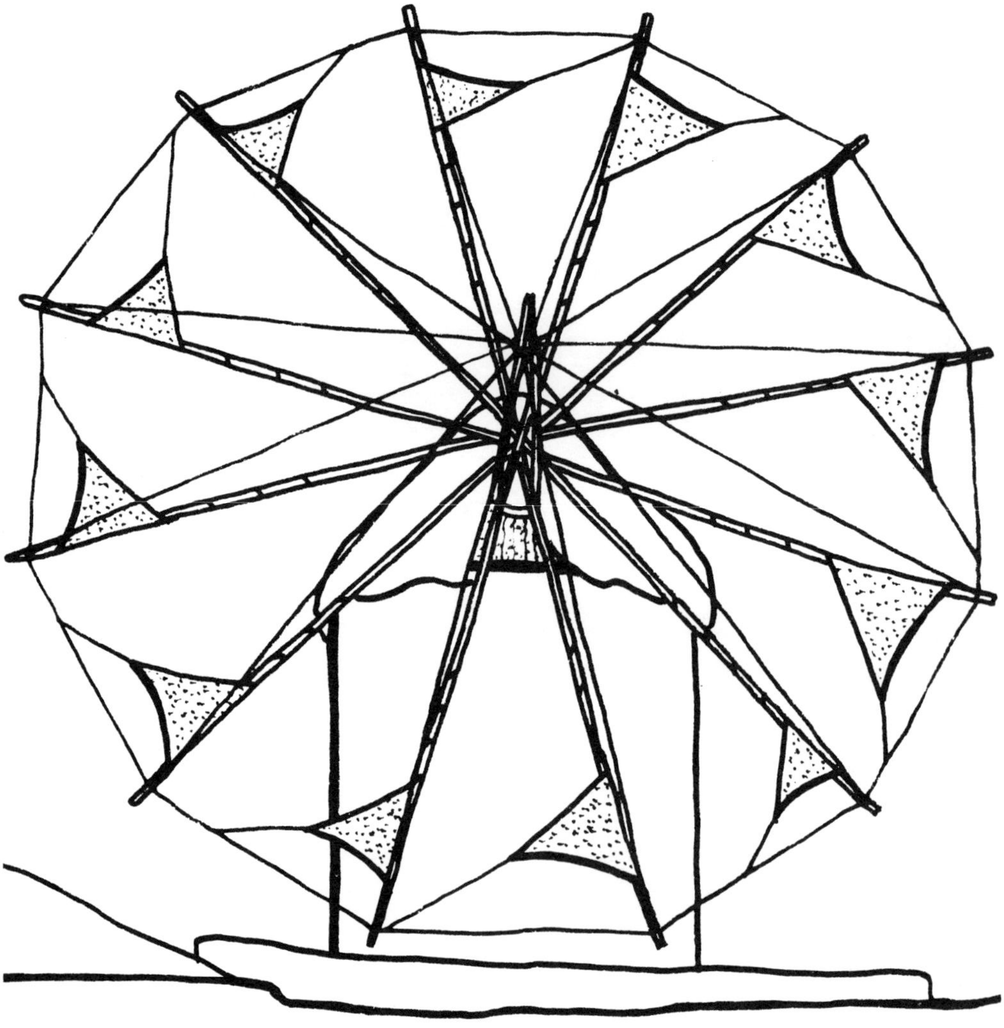

An ancient windmill located on the island of Mykonos in Greece and still working today. This grinding mill is constructed entirely of local materials: the sails are canvas, the spars wood, the roof thatch, and the tower stone. The orientation sails are set manually to suit the wind conditions.

VI. Wind Systems and Design

Wind energy systems are made up of several basic components fitted together in a specific way. The working system may be suited to a specific energy need or to more general energy needs. The combination of the various components determines the overall efficiency and use for the wind system.

The components necessary for any wind system are a collector, means of support, means of orientation, control system, conversion system, distribution system, energy storage, energy use, auxiliary system, maintenance system, and safety devices. All of these components are critical to the proper design and performance of a reliable wind collector system. (See Figure 19.)

The *collector* provides a means of collecting the wind and making it more useful. The wind collector usually changes the airflow of the wind into more specific motion (such as rotating motion or up-and-down motion). An *orientation* device keeps the wind collector facing properly into the ever changing wind. The wind collector is connected to a *conversion* system, which changes the motion of the wind collector into a readily usable form of energy. Some means of *support* holds the wind collector and converter up in the air. The converter is connected to a *distribution* system to deliver the energy to its places of use. Interconnected with the distribution system is an energy *storage* system, which saves excess energy for later use. Different *energy uses* are interconnected with the distribution and storage systems, providing places for the energy produced by the wind system to be utilized. An *auxiliary* energy system is interconnected with the distribution and energy storage system to provide backup energy when the wind system is not working or when the energy storage is empty. A regular *maintenance* system is provided to keep the components in good order and to insure proper performance from the wind energy system. A *safety* system prevents the wind collector system from damaging other things or beings and

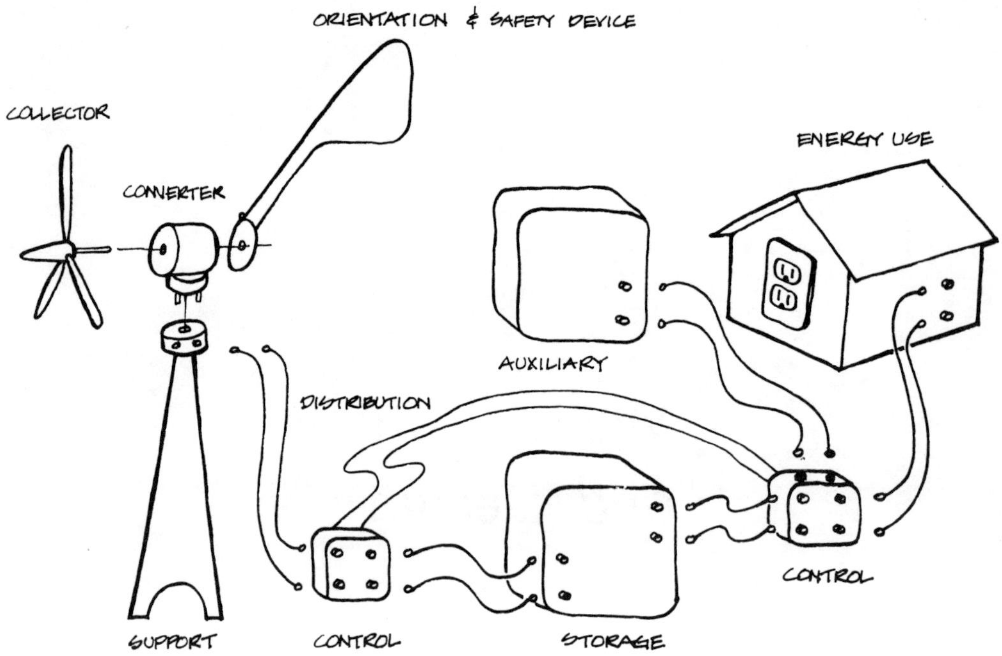

Figure 19. Typical wind system components for domestic electricity generation.

from being damaged by severe wind or weather conditions. The safety system should provide for automatic shutdown of the wind energy system in the event of an emergency. Finally, the *control* system monitors and adjusts the components to facilitate proper performance from the complete wind energy system. The control system interconnects all wind system components. The many separate components comprising the wind energy system are made to work as a responsive and unified system by the control subsystem. The operating modes of a basic wind system are shown in Figure 20.

It is useful to examine some specific wind energy systems to see what their various components are. A window is a simple yet sophisticated wind energy system that is familiar to most people. The window opening is the collector; the orientation of the window is fixed by the wall it is mounted in; the window itself is the converter; the wall in which the window is mounted is the means of support; air in the room inside provides for distribution; energy storage is provided by the wind itself, which is constantly available for use; ventilation, cooling, and drying are the energy uses; auxiliary systems include window fans, hand-held fans, and air blowers; maintenance is the normal operation and repair of the window; safety is assured by devices such as sash weights that prevent the window from being inadvertently opened or closed; and the control system is the user.

A well-known wind energy system is the traditional Dutch windmill. Four blades made carefully of a wood lattice covered with canvas form the wind collector; a manually operated jack and winch orient the machine to the wind; a series of wooden shafts and gears converts the wind collector's rotational motion

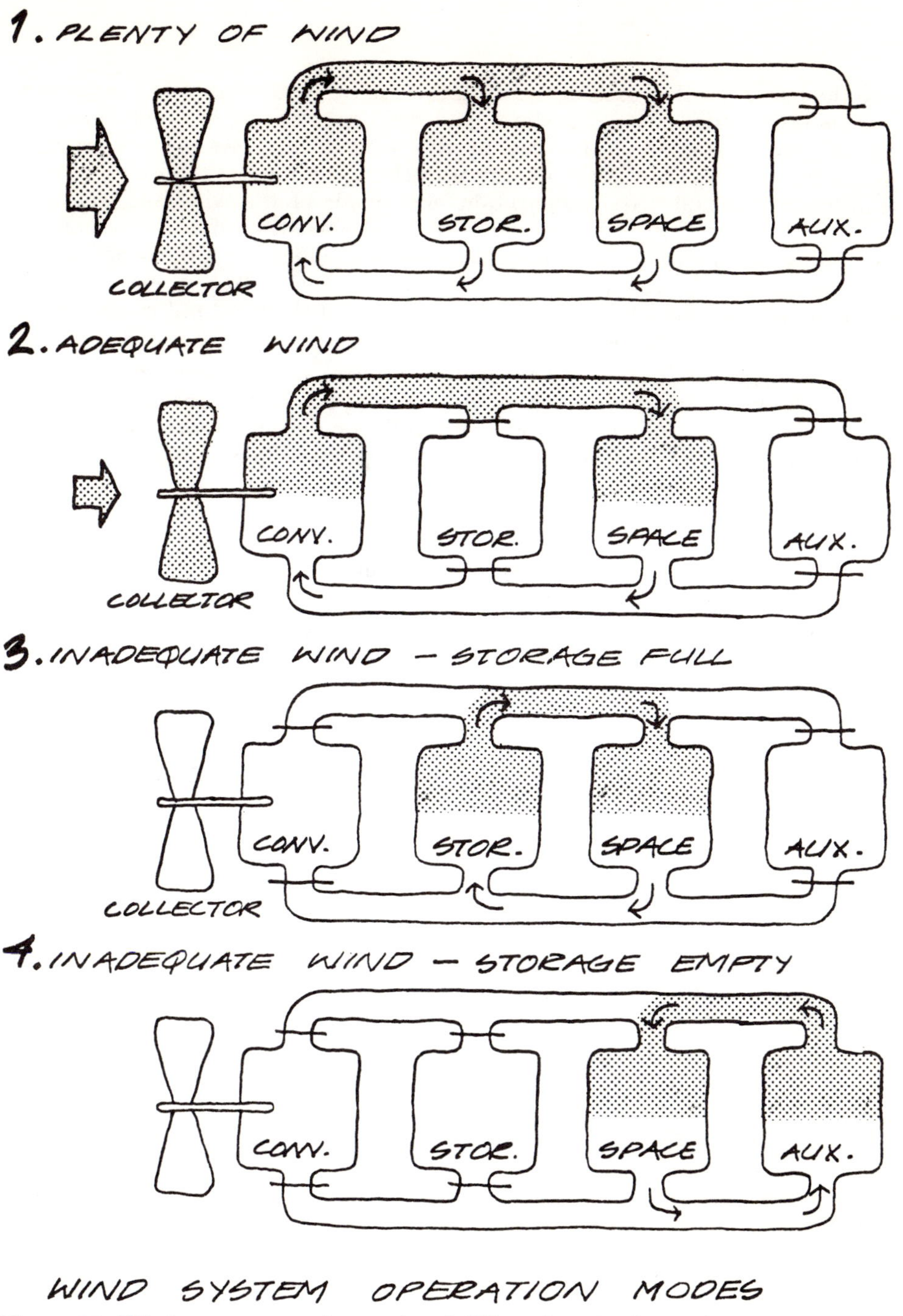

Figure 20. Wind system operating modes: **1.** When there is plenty of wind the system can supply needed power and the excess can be stored. **2.** When the wind is adequate to supply only the needed power, the storage system is disconnected. **3.** When the wind speed drops below that necessary for power generation, power is drawn from storage and the collector system is disconnected. **4.** When the stored power is depleted, an auxiliary system must be used; these include diesel generators and utility electricity networks.

into the faster rotation used in the mill; the windmill structure itself is the support; another series of wooden shafts and gears provides for the distribution of energy; energy storage is provided by the wind itself; powering the grinding stones for grain milling is the energy use; there is no direct auxiliary system; maintenance includes daily operation and repair of the windmill; safety is insured by a large manually operated drum brake on the main turbine shaft, by the jacking mechanism, which moves the rotating blades out of the wind, and by the slow rotational speed of the blades themselves; the control system is the mill operator, who adjusts the area of canvas spread on the windmill blades (more canvas for low winds and less canvas for high winds).

The Dutch windmill is an excellent example of an integrated wind system. It is much like the window, which provides a view, lets in sunlight, gives ventilation, and provides cooling. A Dutch windmill not only supports the wind collector system and its components, but provides a protected enclosure for the wind system components and a place for milling grain, packaging flour, storing grain and flour, and selling flour. The miller and his family might even live in the windmill if it is large enough. A wind collector system of this type is not overwhelmingly expensive because it has so many uses and does so much more than just collect and convert wind energy.

A wind electric system is a bit more complicated and sophisticated wind energy system. The collector is a turbine, usually with two or three blades, that rotates with the wind; tail vanes or smaller rotors are used to orient the device into the wind; a generator is the converter, which changes the rotating motion of the turbine to electricity; a tower of metal or wood provides support for the wind turbine and generator; distribution of the electricity is provided by a system of wires; energy storage is provided by batteries like those in an automobile or bus; electricity consumption is the energy use; auxiliary energy may be provided by the local utility company or by a gasoline-powered generator; maintenance includes the daily inspection, monthly check and repair of the system, and periodic replacement of worn parts; safety is provided by a fail-safe shutoff system, which automatically turns the wind collector out of the wind whenever anything is not working in a normal manner; and the control system is an electronic device that compares energy needs with wind collector energy output and energy storage capacity and adjusts the system accordingly.

The initial cost of any active wind system is relatively large. It is necessary to have a minimum collection area to use the wind at all. The area must be large because of the diffuse nature of the wind, and thus, the collector system has a high initial cost. Yet, the maintenance, operating, and social costs over the lifetime of the wind system make it less expensive than more conventional energy sources such as uranium, oil, electricity, natural gas, or coal. Such conventional systems have relatively low initial costs, but much higher and somewhat unpredictable operating costs. It has become increasingly important to examine the overall life-cycle costs for the various energy systems we use now and will use in the future.

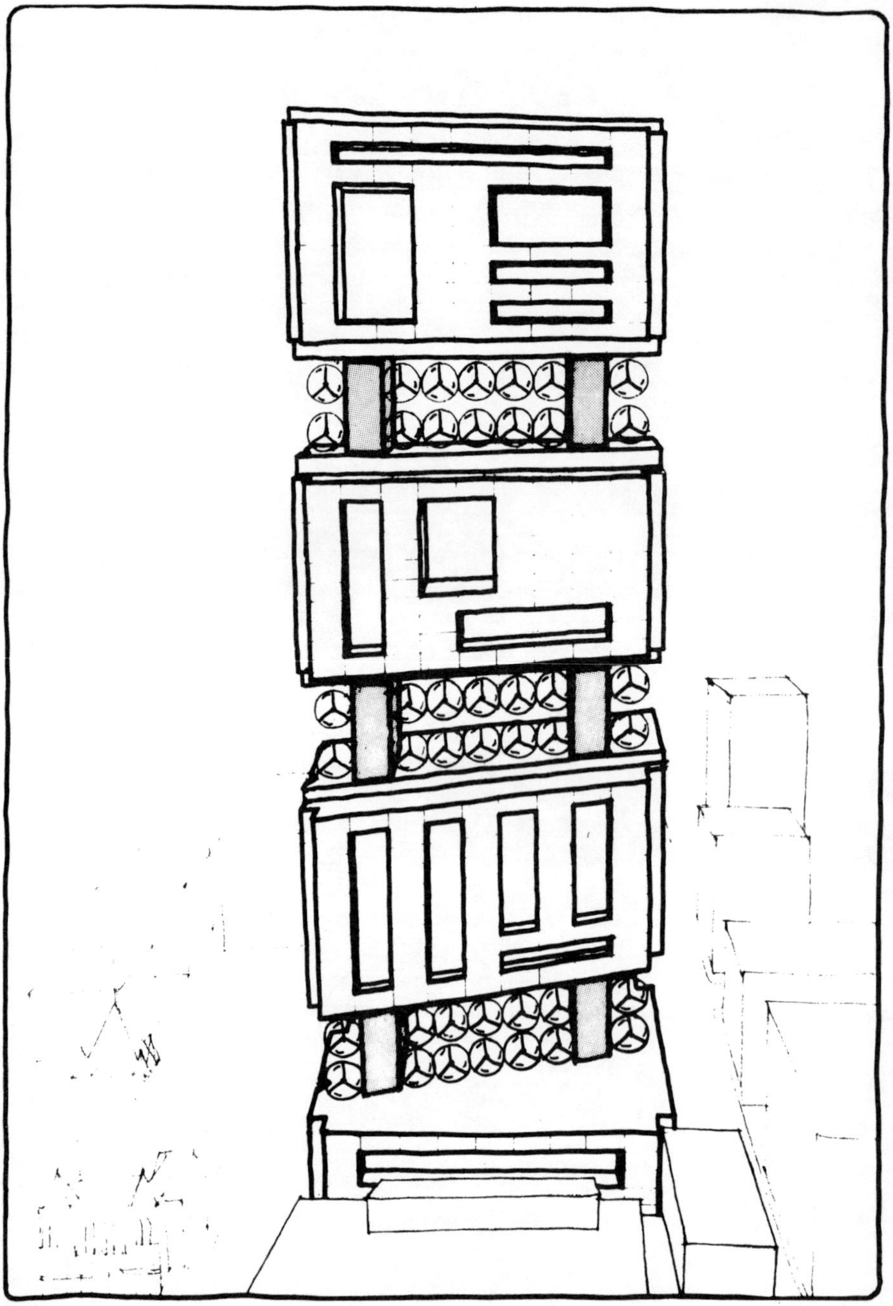

A wind system utilization study for the John Hancock Building in Boston. The design retains the building's original structure, reduces the floor area by one third, and uses existing elevator and mechanical shafts. Two-stage wind collectors are proposed for the openings, one wind turbine in front of another. The installation could supply the electrical power for the Hancock Center and the eight surrounding blocks.

VII. Building Forms

We saw in Chapter II that buildings and structures disturb the wind as it sweeps past them. The amount of disturbance depends on the size and shape of a building and its location relative to other buildings and structures nearby. Almost anything solid in the wind's path presents a barrier to the flow of air. The wind must flow around the object since it cannot go through it. Molecules in the air must crowd closer together because the volume they occupy is lessened by the building's presence. Some molecules must take longer paths than others; they no longer travel in a straight line. However, all the molecules must keep up with one another, so those that must travel farther need to move faster than before, and these differences in speed cause differences in pressure.

These phenomena all contribute to the bothersome winds in city streets and the vortices and eddies at the corners of buildings, over their rooftops, and near their bases, as shown in Figure 21. They also increase the heat loss from the building and cause additional wind problems for other buildings and structures. Ultimately, people too are affected by these wind conditions. Yet, people design and build structures to provide useful shelter and not to cause further problems.

It is difficult, if not impossible, to design a structure that does not disturb the wind. It is possible, however, to design so that the disturbance is beneficial rather than detrimental. This requires careful planning and understanding of the characteristics of the wind. Careful design and respect for people, trees, and buildings nearby, make it possible to achieve more integrated wind systems similar to the Dutch windmill. Simultaneously, the wind problems normally associated with buildings can be substantially alleviated. It is also possible to design a building that increases the wind velocity to make it more useful. And increasing the useful wind velocity at a wind collector's location increases the energy output potential dramatically. How much is "dramatically"? Current tests and installations indicate between two and 15 times! Figure 24 shows that the wind's velocity can increase by a factor of two or more when a diffuser is added to a rooftop. Since the power in the wind is proportional to the velocity cubed, the power can easily increase eightfold. The cost per unit of energy (kWh) is greatly reduced, making wind-generated energy more competitive eco-

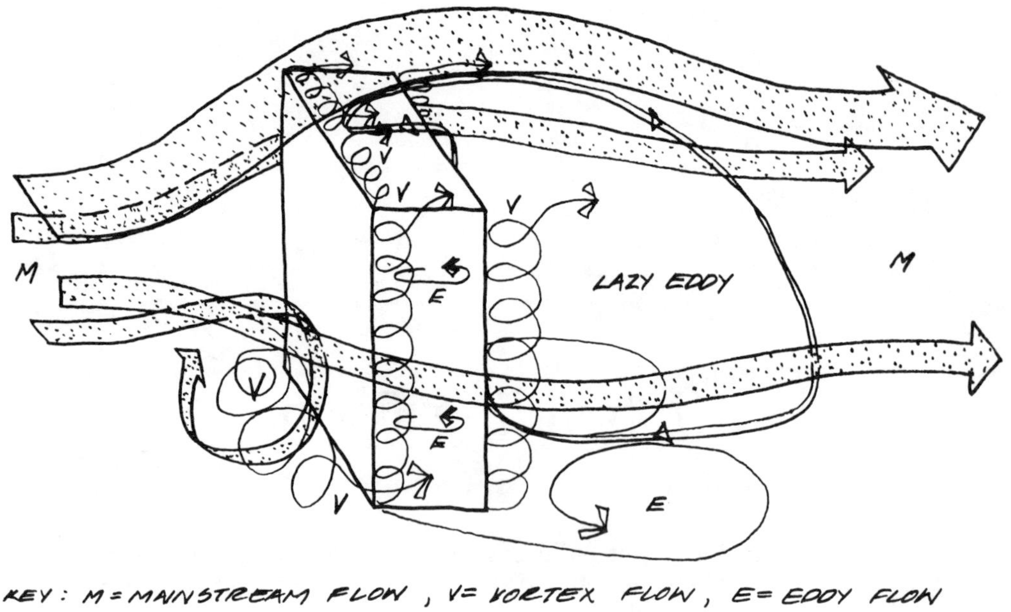

Figure 21. Wind flow around a solid structure. Wind is diverted toward the bottom of the structure, over its top, and around the sides. Notice what happens at the corners and on the leeward side of the structure. (Adapted from *Wind Effects Around Buildings* by Richard Aynsley.)

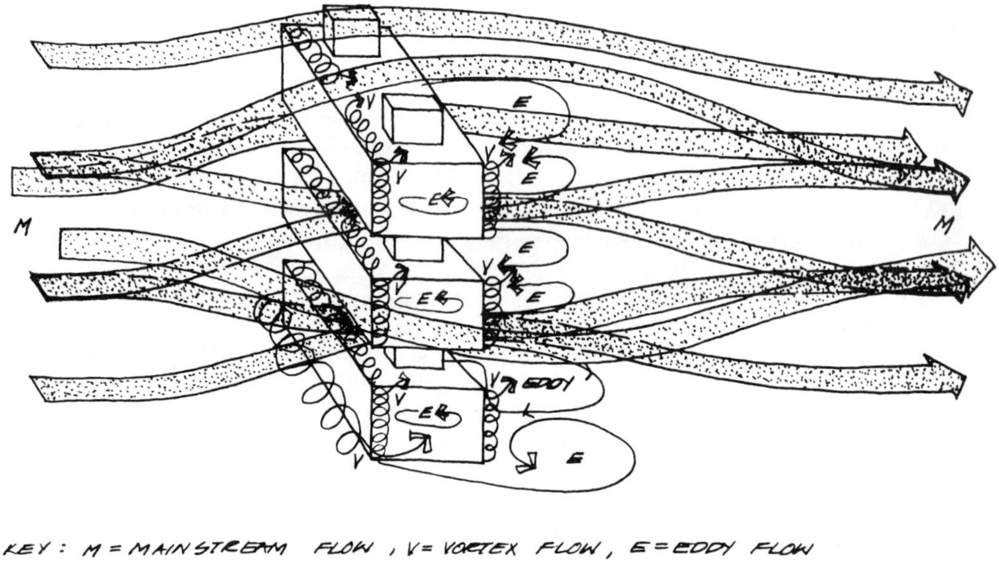

Figure 22. Wind flows around and through a structure with openings, which reduces the disturbance around the building.

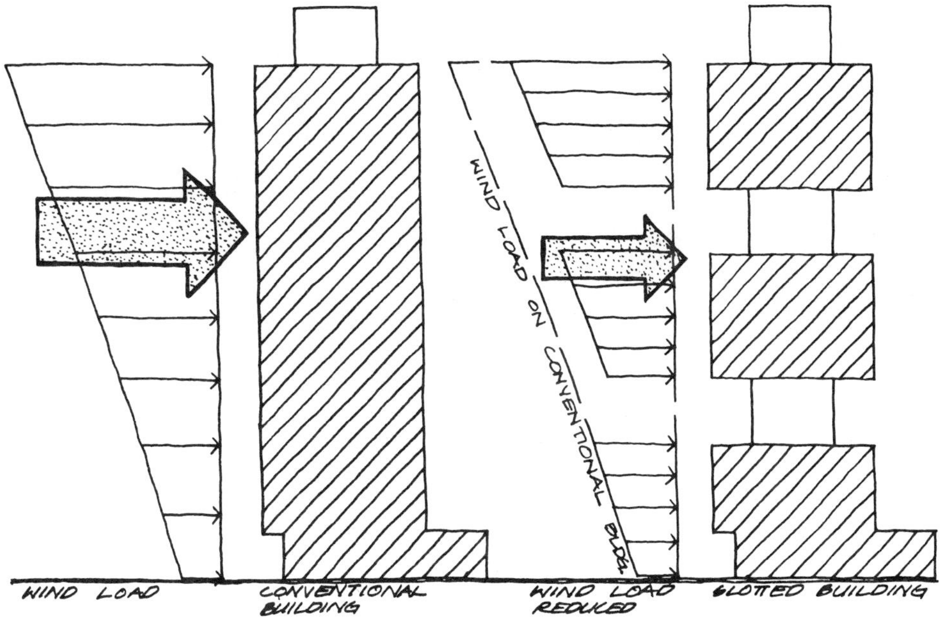

Figure 23. The wind load on a slotted building is considerably smaller than that on a conventional building.

nomically with conventional energy. In some areas with high wind speeds, wind-generated energy may even cost much less than conventional energy.

Generally, a building with many holes or openings through it has a reasonably low resistance to the wind. Something similar to a snow fence or picket fence would work very well. Yet, it is quite difficult to construct buildings with so many openings. It is reasonable and very useful to provide many operable windows for natural ventilation and cooling during the warm seasons. During the cool seasons it is necessary to close most (if not all) of the operable windows to save energy for heating. Ventilation is still necessary, especially during the cool season, to prevent stuffiness and to maintain comfort levels. The ventilation is usually provided by fresh air inlets into the heating system or through energy-conserving heat exchangers feeding directly into the building. A conventional building with operable windows has a greater wind resistance during the cool seasons than during the warm ones. Actually, this is just the opposite of what is most desirable because the wind is usually stronger during the cool seasons than during the warm seasons!

A low-wind-resistance building needs to maintain a low resistance profile throughout the year, especially during the seasons with the strongest winds. The openings that let the wind pass through must be integrated parts of the building and separate from any operable windows for natural ventilation or cooling. These openings work well when they are distributed evenly over the exposed face of a building, allowing the wind to pass evenly through as shown in Figures, 22 and 23. Wind collectors may be placed within these openings; much of the

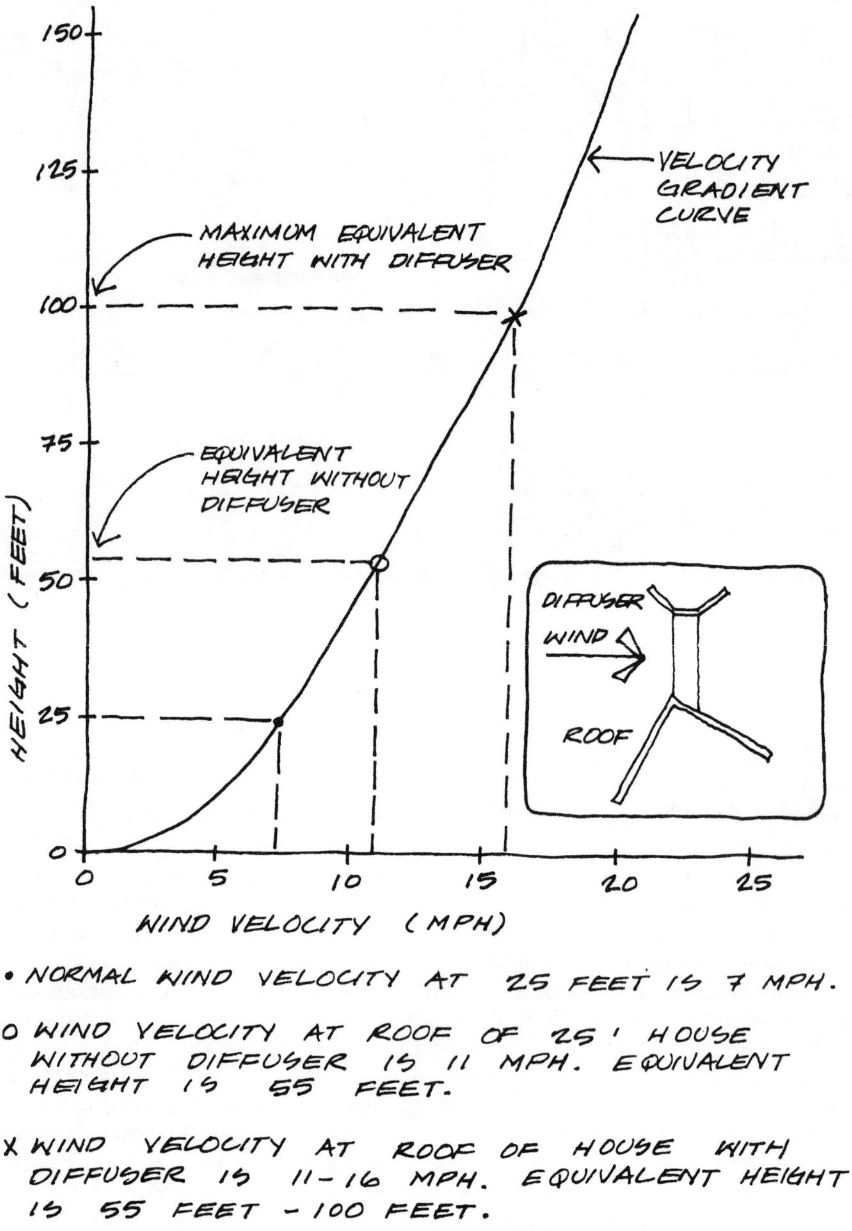

Figure 24. A rooftop diffuser can substantially increase the amount of energy extractable from the wind. The wind velocity normally increases with height, as indicated by the velocity gradient curve. In this example the wind velocity at a height of 25 feet is 7 mph. Since a 25-foot-high building partially obstructs the wind, the wind velocity over the top of the building increases to 11 mph. This speed is normally found at a height of 55 feet (equivalent height). Adding a diffuser to the roof can cause the wind speed to increase to 16 mph, which is normally found at a height of 100 feet.

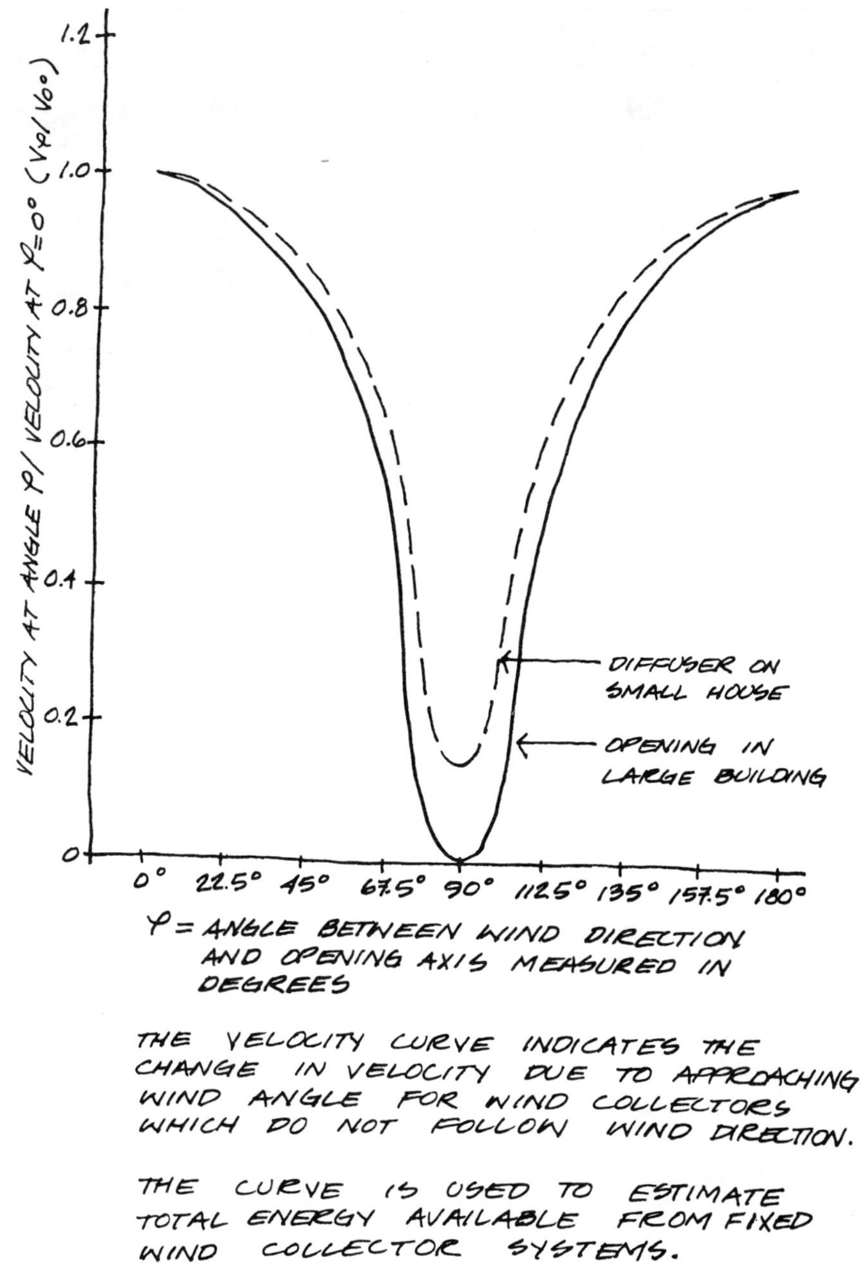

Figure 25. Wind velocity at an opening in a structure varies with the angle of the approaching wind. The velocity is greatest when the wind approaches the opening head on (the angle $\varphi = 0$), and decreases to zero when the wind approaches at a right angle to the axis of the opening. The decrease in availabe energy is not so great in the case of a diffuser.

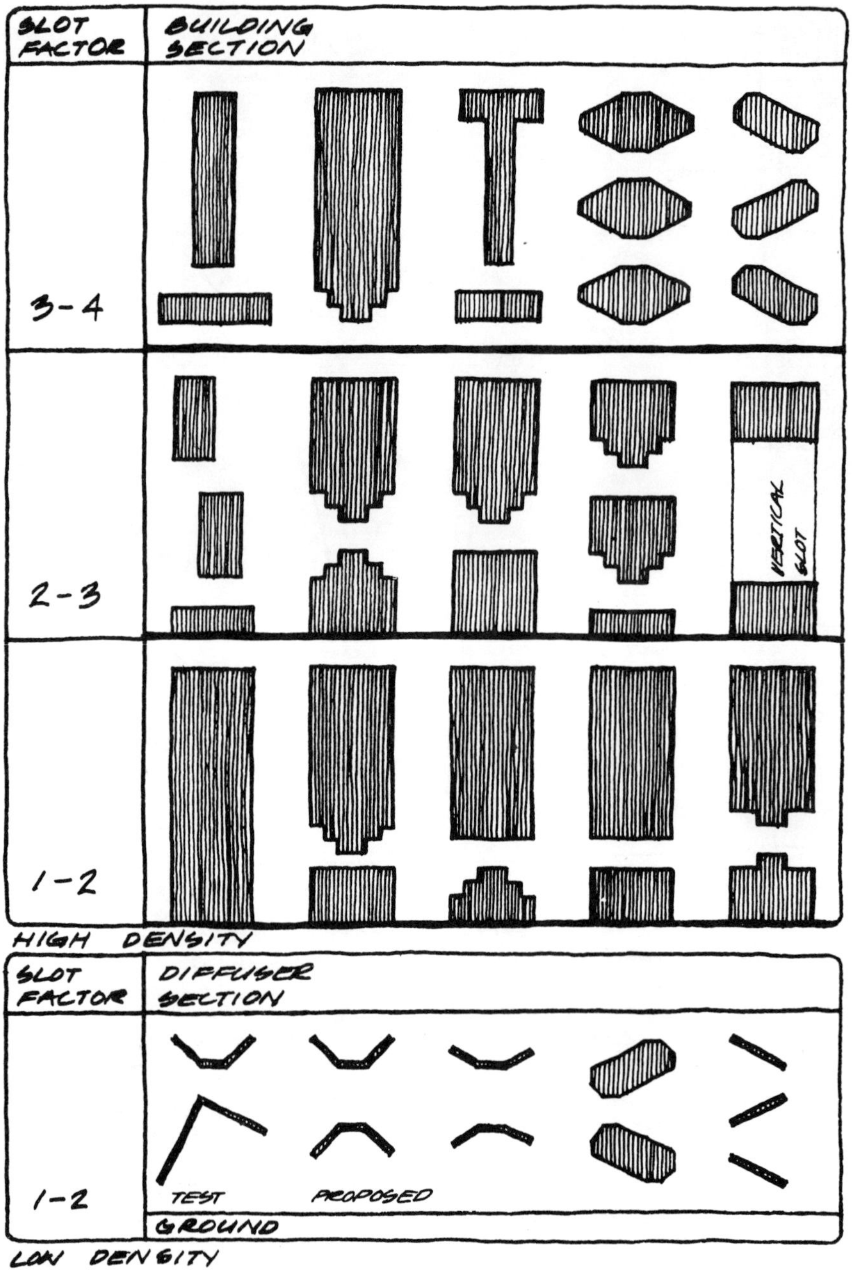

Figure 26. Wind velocity increase capabilities for buildings of different shapes. High-density buildings include large office buildings and high rises; low-density structures include homes and small buildings.

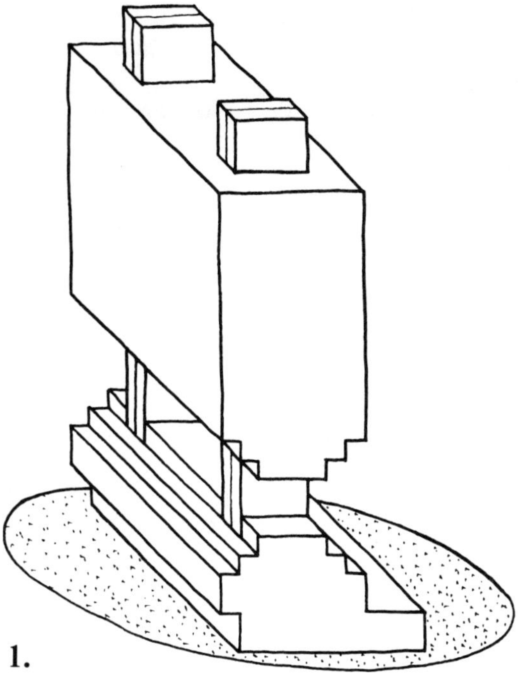

1.

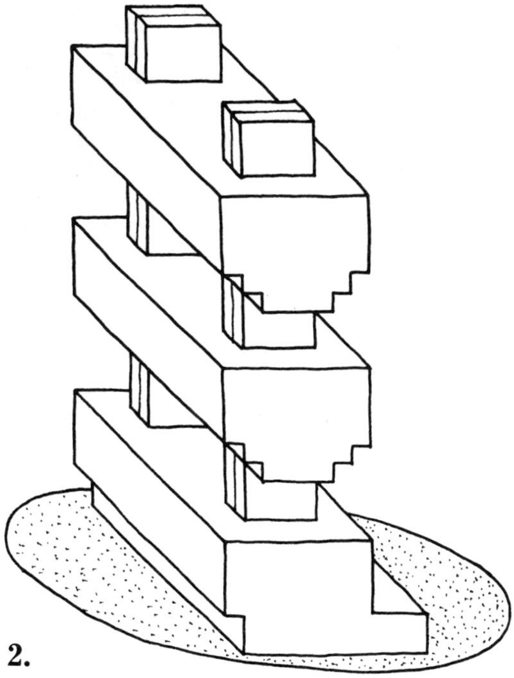

2.

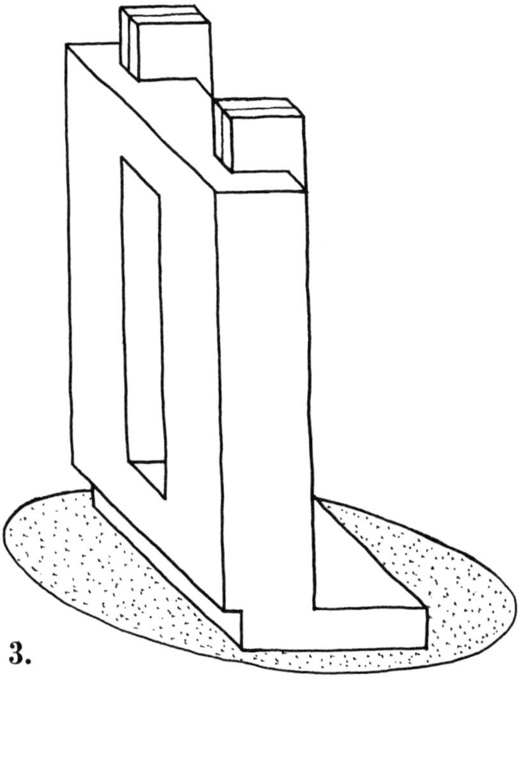

3.

Figure 27. Possible wind duct configurations for buildings: **1.** one tapered duct close to the ground; **2.** two partially tapered ducts spaced evenly in the face of the building; **3.** one large vertical duct in a building resembling the twin towers of the World Trade Center in New York City.

wind will still pass through, which reduces the wind load, and substantial amounts of energy can be produced. (See Figure 29 in Chapter VIII and the illustration at the beginning of this chapter.)

The orientation of the building and the type of wind collector used are important considerations for this wind energy arrangement. A building is stationary and cannot orient itself into the wind as a conventional wind collector can. The optimum building orientation is that with the greatest wind-energy-gathering potential (remember that a wind duct has equal capabilities from both directions). A graphical representation of the change in wind velocity according to the angle between the wind and the axis of the duct opening is shown in Figure 25.

Some shapes or forms work well in reducing wind problems caused by most buildings. Other building forms work well in one or two primary directions to increase wind velocities and to reduce common wind problems as well. Still other building forms work well in increasing wind velocities from many directions while reducing the other wind problems at the same time. Some of these forms are shown in Figure 26 for both low-density (single family) and high-density (multiple family or commercial) buildings. The relative performance of different building forms is shown by the wind velocity multiplier given for different groups of forms. Multiplying the normal wind velocity by the velocity multiplier gives the relative wind velocity in that particular wind duct location. This velocity varies markedly from the wind velocity that would normally be found at that location if the building were not there. With this information and the variation in velocity at the duct with change in wind direction (Figure 25), the local wind data, and the energy calculation chart, it is possible to determine the energy available from such an arrangement. The energy capabilities of these integrated wind ducts are usually substantial. Yet, they are not to be obtained without careful consideration for safety, public acceptance, construction costs, and wind rights.

Several design possibilities for wind systems integrated with building forms are shown in this book. These designs do not include all the alternatives, but only begin to explore the potentials for utilizing the wind without changing existing wind patterns drastically or placing unnecessary burdens upon other energy systems or ecosystems in the process.

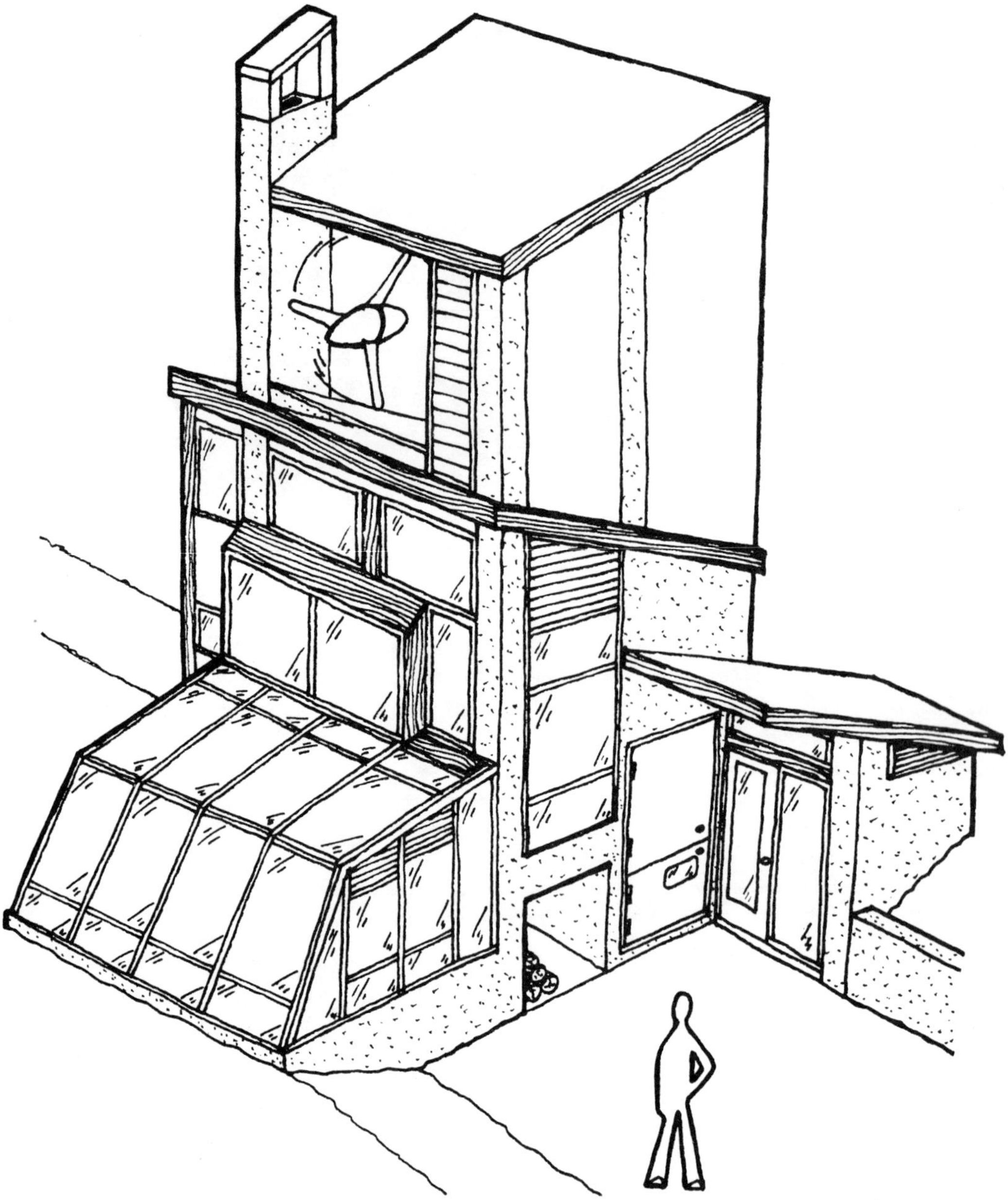

A design for an integrated energy shelter that uses passive and active wind systems, solar heating, rainwater collection, a greenhouse, and food and waste recycling. This shelter can provide all the energy needed by a single person.

VIII. Wind and Other Energy Sources

Wind energy is just one of many alternative energy sources being considered for further development and use. *Alternative* energy sources are those not commonly used as major energy sources in the world, and they are characterized by a short-term renewability. Alternative energy sources include solar energy, wind energy, wood energy, water power, rainwater collection, geothermal energy, waste recycling, wave and tidal energy, biogas conversion, hydrogen gas, methanol, and "people power." *Conventional* energy sources are those commonly used as major energy sources, and they are characterized by long-term renewability. Conventional energy sources include gasoline, fuel oil, natural gas, coal, chemical and hydroelectric power, and nuclear fission.

Alternative energy sources were the common primary energy sources prior to the current fossil fuel age! Many of these "new" alternatives are not really new at all. They are often revivals of well-developed energy sources of the past, modified to be compatible with current manufacturing and control systems. Much can be learned from the energy sources of the past about the simplicity, sophistication, ecology, and integration of systems. There has been a trend toward more complex, high-energy-consuming systems (air conditioners, for instance) instead of less complex, less-energy-consuming systems (windows that open and close) for the same problem. This trend is slowly changing in response to an increased awareness of the immense waste of energy, material, and effort that has resulted. The United States of America, in particular, has contributed greatly to the overconsumption trend. It is now up to the United States to set an example for other nations by reversing that trend and by launching a more ecologically sound policy that harmonizes with nature.

Wind is a special product of solar energy. The wind is constantly renewed by the sun's differential heating of the earth and its surrounding atmosphere.

Actually, most energy sources are forms of solar energy, either directly (sunlight coming through windows) or indirectly (wood growth). Fossil fuels are made from the decayed and compressed remains of plant and animal matter. Hydroelectric power is provided by the continuous evaporation of water into the atmosphere and its redistribution as rain. Nuclear energy is one form of energy not directly linked to the sun, but systems relying on solar energy are used to control, contain, and utilize it. Most energy sources are renewable. Some require longer natural renewal periods than others. The natural period of renewal determines how quickly the energy source may be utilized without upsetting the critical balance of nature. Quickly renewable energy sources (such as direct solar radiation, waves, and wind) may be more fully utilized because they are renewed continuously. More slowly renewable energy sources (such as wood, water power, biogas, waste recycling, rainwater collection, geothermal energy, and hydrogen gas) must be carefully conserved because they are renewed on a longer-term basis of weeks, months, or years. Very slowly renewable energy sources (such as coal, natural gas, oil, and fissionable materials) must be the most carefully conserved because they are renewed only over extremely long time periods and only under certain conditions.

Wind is quite compatible with most conventional and alternative energy sources. Mutually beneficial relationships exist between certain energy sources, which allow them to be used easily in conjunction with one another. One excellent example of this type of relationship exists between wind energy and direct solar energy. During the summer, when the sun is strongest, the wind is weakest. And during the winter, when the sun is weakest, the wind is strongest. Over a shorter period of time, the wind is usually strongest when the sun is not out and the weather is cloudy; the wind is usually weakest when the sun is shining brightly and weather is clear. What about wind and wood? Or wind and biogas? Or even wind and people? Wind needs help from time to time to insure a constant and reliable energy supply.

It becomes apparent that no single form of energy is completely reliable. Even "reliable" energy sources such as coal-powered electric utilities require periodic maintenance, are subject to emergency shutdown under extreme weather conditions, and undergo fluctuations in fuel supplies. So it is only through the careful combination and integration of several different energy sources that a reasonably reliable energy supply can be maintained. There are many ways to combine different energy sources into one integrated energy system. Some people have realized this important notion of integrated energy and have produced very reliable energy supplies. The Ecol-Operation at MacDonald College in Canada, the New Alchemy Institute's Ark on Cape Cod, Street Farm in England, and the Alternate Energy Center at Ramapo College (Figure 28) are some examples of working integrated energy systems. Research is currently underway to design and build more integrated energy systems. And many practical applications of integrated systems are being made by homeowners, do-it-yourselfers, and students who are interested in more ecological and reliable energy sources.

An especially interesting project is now underway near Beallsville, Ohio. The project, called "Raven Rocks" after the birds that inhabit the area, was created by an environmentally conscious group of people who united to purchase many acres of land in the midst of an area destined to be strip-mined for coal. Raven Rocks will be an integrated energy community completely separated

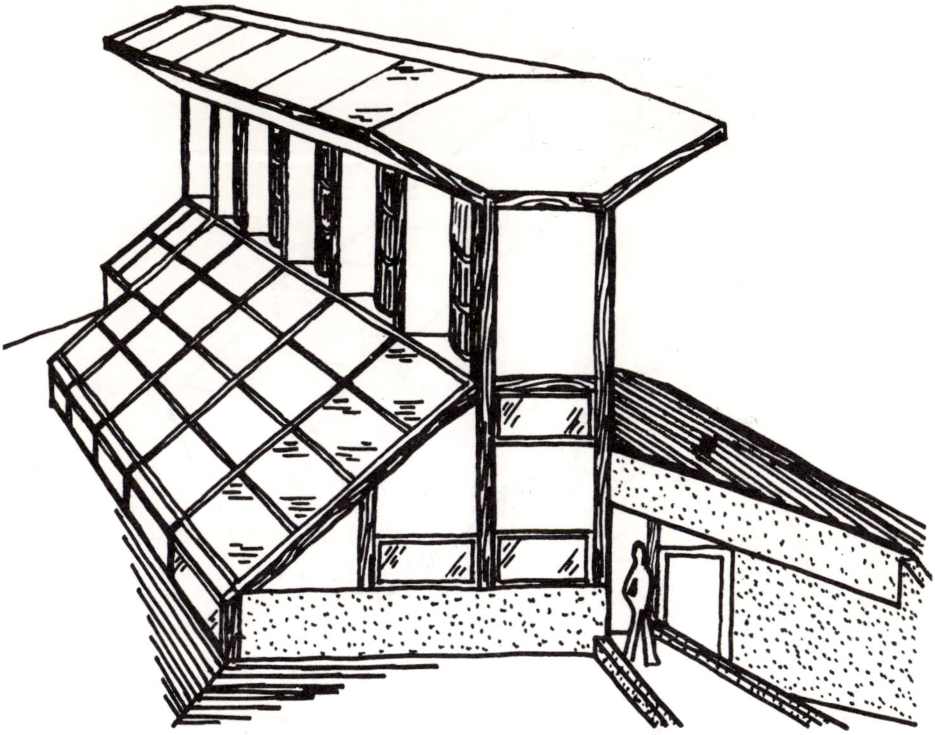

Figure 28. The Alternate Energy Center at Ramapo College in Mahwah, New Jersey, uses ducted wind collectors, direct solar heating, and wood for all its energy needs. The center is to be used for aquaculture research, laboratory work, classes, and demonstrations.

from public utilities. The project will use solar space and water heating, a year-round greenhouse for food production, a wind duct for wind-generated electricity, wind-generated hydrogen gas for cooking, wood auxiliary heating, composting dry toilets, waste recycling, rainwater collection, and an earth-covered building. It will provide a home for six people and community activities as well.

A project on a much smaller scale, yet striving for the same energy independence (without sacrificing comfort or enjoyment), is illustrated at the beginning of this chapter. This project was designed by Dan Scully, myself, and our students at the University of Massachusetts-Amherst summer session on Integrated Energy Systems. This Integrated Energy Shelter is designed to provide for the energy and living needs of one person in a practical manner. This design is not suitable for everyone, but it is an example of the integration of many systems into a single coherent structure. The project also demonstrates the relative amounts of space and equipment needed to provide the basic necessities for a single person.

One of the most practical ways to reduce our overall energy needs is by energy conservation and the wise use of the energy sources available each day.

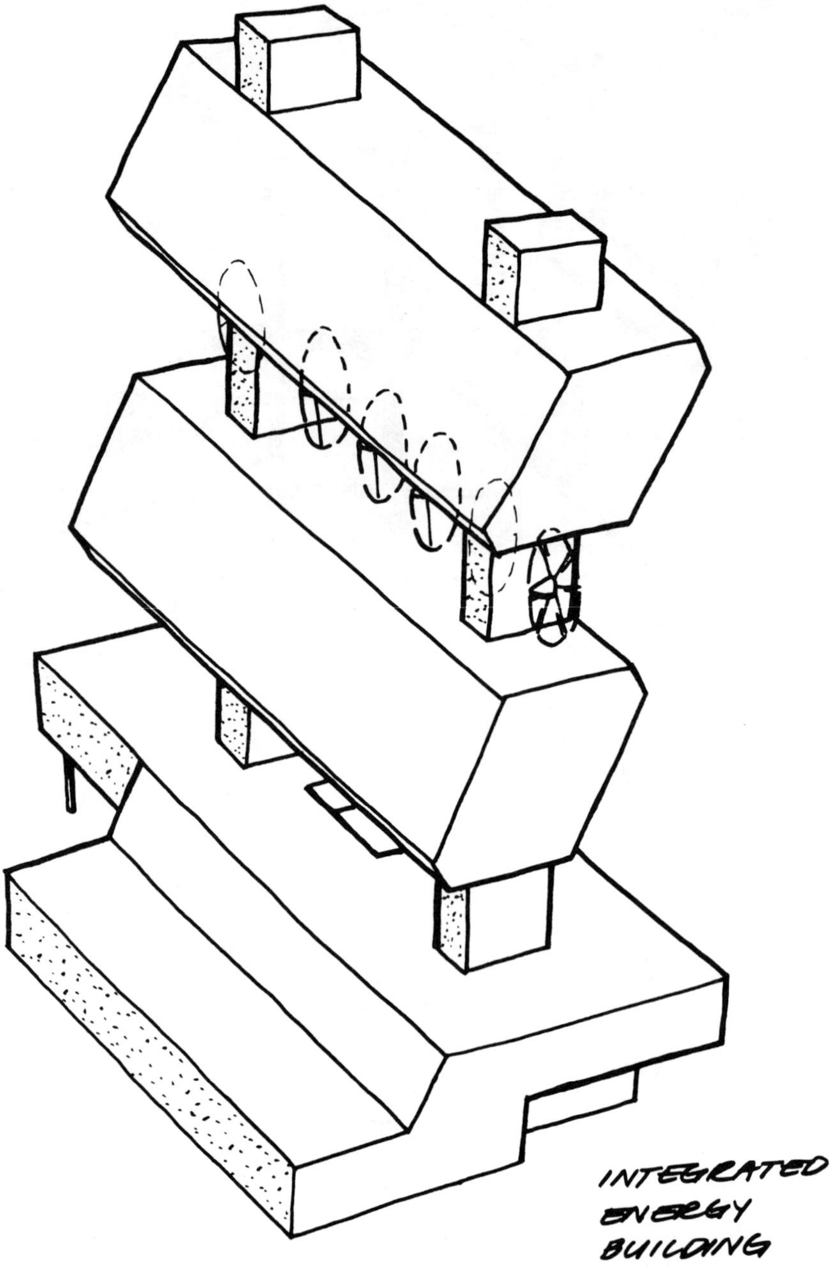

Figure 29. A conceptual design for an integrated energy building. The design incorporates both active and passive wind collectors and solar collectors for space and water heating. In the lower plaza there is space for off-street parking and for commercial uses. The upper portions of the structure are for residential or office use.

Conservation of energy may reduce total energy needs in the United States by as much as 50 percent. Garbage and trash are a huge, yet often neglected, energy source. The use of garbage and trash as a fuel supply for energy systems is not only economical today, but it would alleviate many problems now encountered at local landfills and dumps. Large areas of land would no longer be given over to unsightly piles of garbage generating foul smells. Current garbage and trash production in the United States could provide enough usable energy to heat all of the homes and apartments in the United States. Biogas is produced by the natural decomposition of garbage. This gas can be recovered economically by drilling a well in the dump. This has been done in California.

Recycling glass and metals can also reduce our reliance on the earth's limited natural resources. Automobiles are recycled: Certain parts are reused directly, and others are melted down for steel to make new, more efficient automobiles. Appliances and machinery may be recycled in a similar fashion. Finally, homes are being recycled as more and more people buy and restore older structures rather than build or buy new ones.

It is important to understand that all energy systems (wind, nuclear, or otherwise) ultimately are created by and for people. People conceive, design, build, and operate the systems, and people make mistakes occasionally, even frequently. It is through the ingenuity and resiliance of people that the systems are able to work reliably to provide for the needs of everyone.

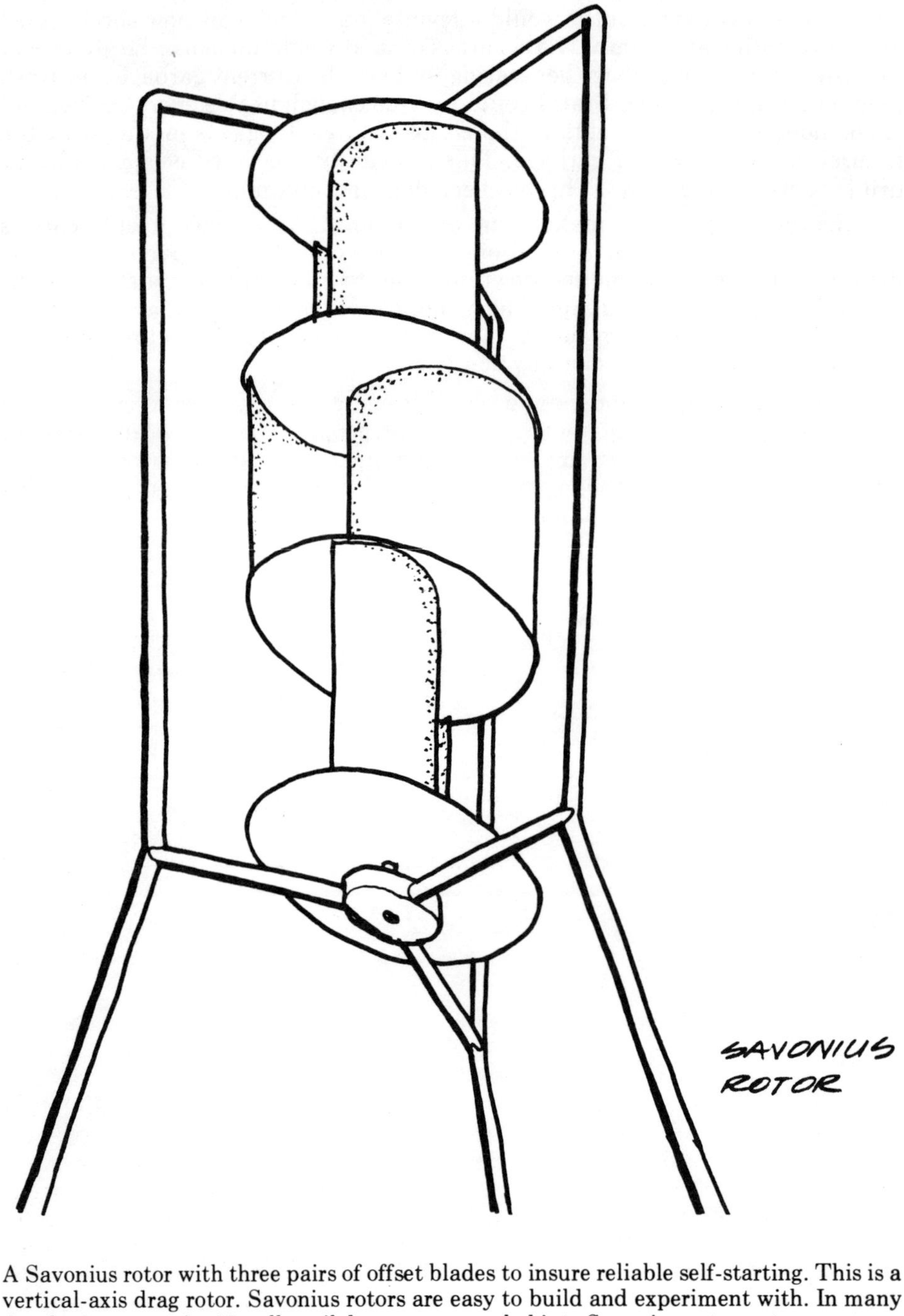

A Savonius rotor with three pairs of offset blades to insure reliable self-starting. This is a vertical-axis drag rotor. Savonius rotors are easy to build and experiment with. In many developing countries 55-gallon oil drums are recycled into Savonius-rotor water pumps.

IX. Putting It All Together, or Designing with Wind

Now it is possible to design a system for the practical use of the wind in most locations and to integrate it with the other energy sources available. There are several questions we must ask: What will the system be used for? What type of system is needed? Where will it be located? How will it be integrated with the building and surroundings? The use for the wind energy, the amount of energy, the type of wind system, the location of the wind system, the type of building, the amount of money available for the wind system, the lay of the land nearby, the type of vegetation, the location of neighboring structures, and the local power company's attitudes and rates are all important factors in the process of "*designing with wind*." Designing with wind is the logical conclusion from all of the information discussed previously and the incorporation of a workable wind system or systems. This is, perhaps, the most difficult part of the entire design process. There are many different needs and variables to satisfy before the system becomes workable.

General guidelines for the design process may help facilitate the design and application of the many factors involved in a wind energy system. A useful diagram is shown in Figure 30. Start with the *uses* for the wind energy because they have an important impact on the design of the entire system. Next, apply the *site* information to determine the size and location of the wind system. Now, apply the *building* needs to see how they may or may not integrate with the wind system uses and the site. Determine, next, whether an *active or passive* system (or both) is appropriate. Then, prepare a *system diagram* showing where the system components might be located and how they might relate to each other. Apply the *system* to the site and to the building to determine how well they work together. This takes us back to the second step (site information) where the process is repeated. When almost all ("all" is a difficult, if not impossible, term

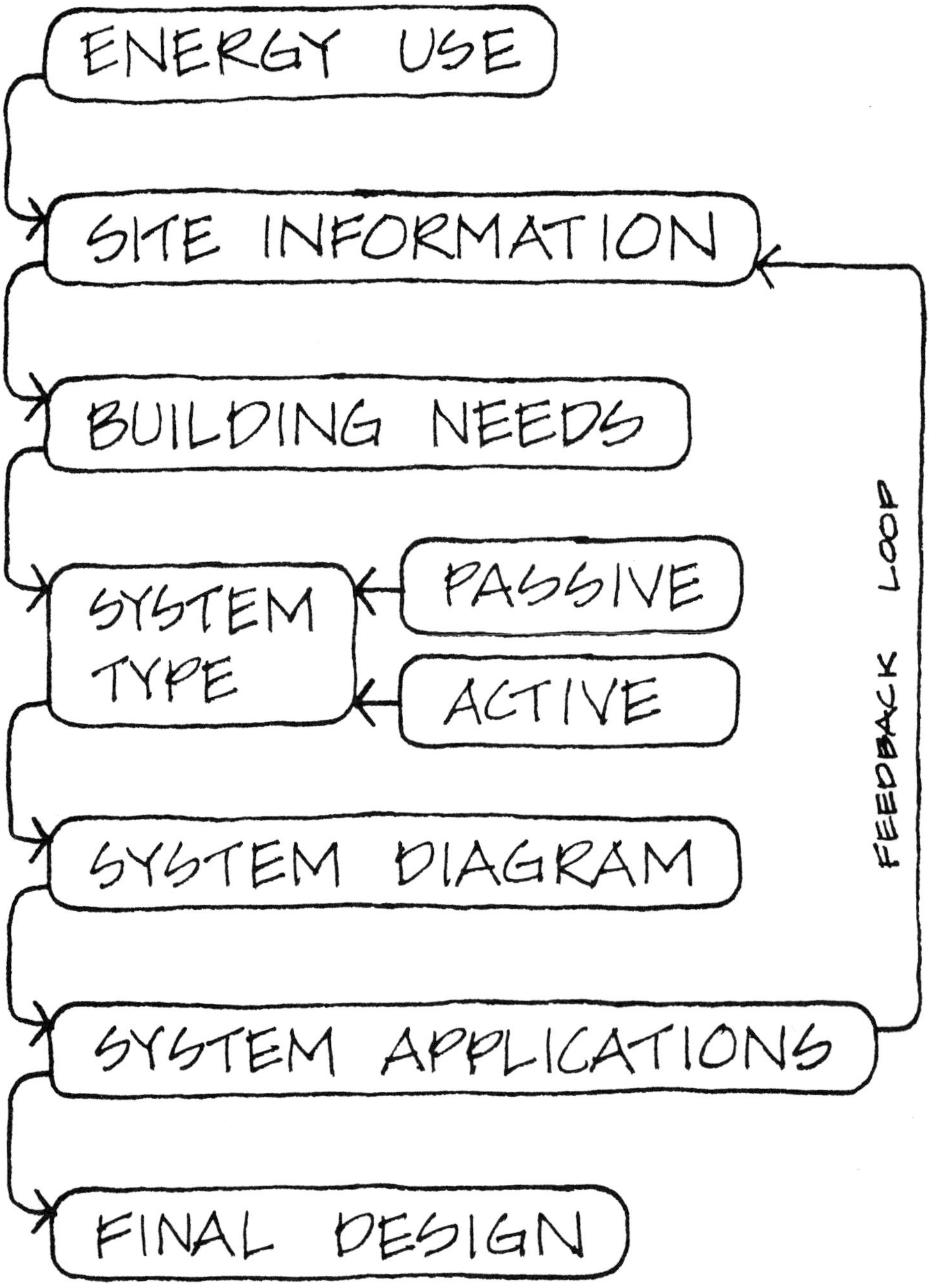

Figure 30. A schematic diagram of the basic steps in designing a wind energy system. There is an important feedback loop between "System applications" and "Site information."

Figure 31. Design for an ecological shelter for Nantasket Beach, near Boston. There is a rooftop wind duct with wind collectors. The top of the wind duct collects and stores rainwater. The active wind system provides all the electricity necessary for the shelter and for a small electric vehicle.

to meet) of the needs and requirements for the wind system are met, the *final design* is completed. The final design might be a page-long list of components to buy and put together, or it may be many pages of drawings and detailed specifications for someone else to put together. The final design depends upon the relative size and complexity of the wind system. A passive household wind system could be a short list of windows and a sketch of their locations; the final design for a commercial building's active wind system could be twenty pages of drawings and a small book of system details and specifications.

Examples of "putting it all together" are shown in Figures 28, 31, and 33. There are many different ways of designing with wind, and these are just a few examples to indicate the wide range of possibilities. Generally, the most cost-

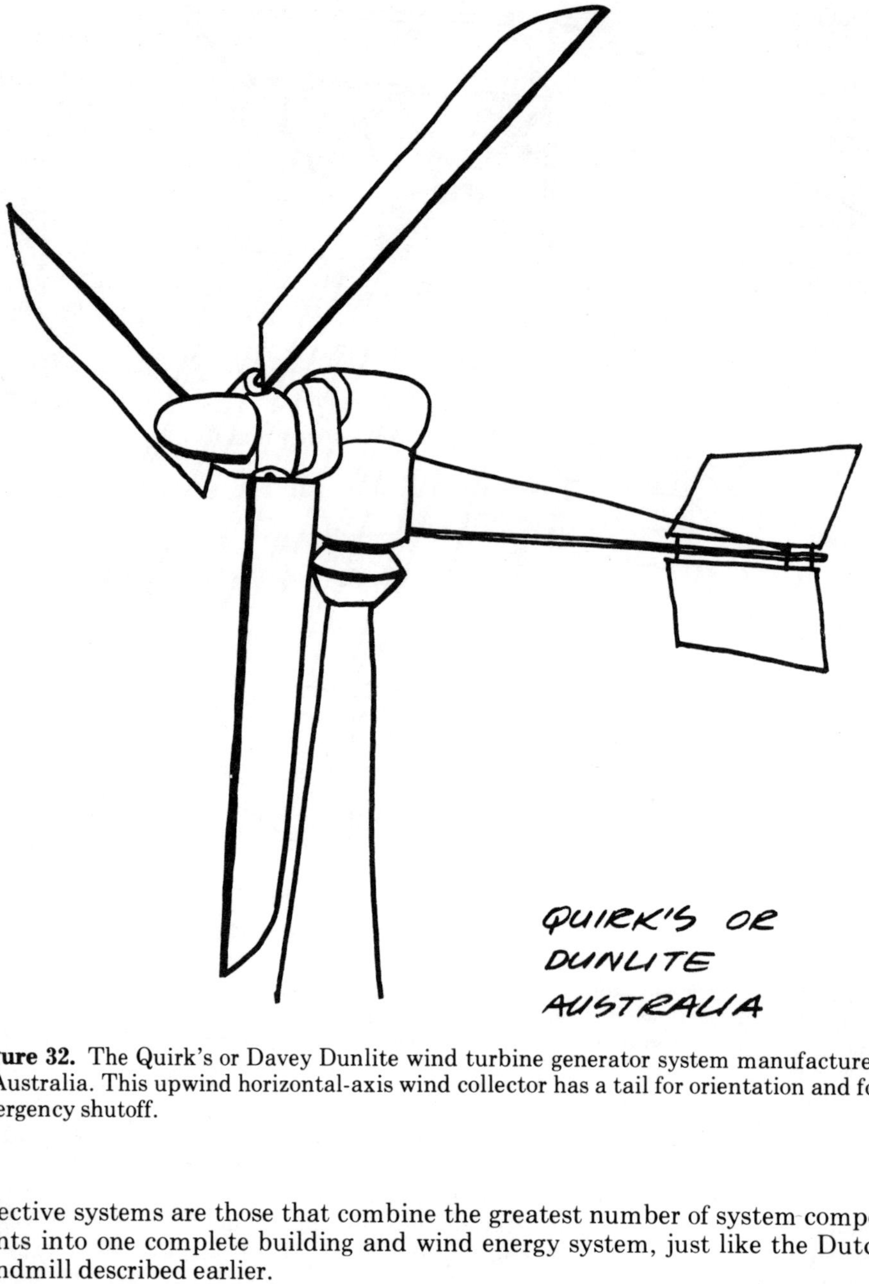

Figure 32. The Quirk's or Davey Dunlite wind turbine generator system manufactured in Australia. This upwind horizontal-axis wind collector has a tail for orientation and for emergency shutoff.

effective systems are those that combine the greatest number of system components into one complete building and wind energy system, just like the Dutch windmill described earlier.

Doing It Yourself

There are many wind systems and components that may be purchased directly from manufacturers or distributors of wind equipment. Most people purchasing wind systems choose an existing system and have it installed by a local distributor. Doing it yourself is yet another way to acquire a wind system.

Many excellent wind system components and kits are available for people wishing to do it themselves. And substantial savings can be realized by carefully designing and building a do-it-yourself wind system from the available components, or even by making the components by hand! The most important requirement for a do-it-yourself wind system is time. Lots of time and effort are needed to accomplish the tasks usually performed by specialized workers in a manufacturing plant. It is quite useful (and sometimes very necessary) to know of local people specializing in engineering, electronics, construction, airflow, and structural design. These people are of great help when a problem or question comes up regarding a specific part of the wind system or its installation.

Perhaps the most important thing to remember when designing and building a wind energy system is *common sense.* Utilization of local and readily available materials from reputable distributors is one example. Keeping the complexity and scope of the wind energy system within reasonable limits is another example. Don't be afraid to ask or to go to someone else for help when it

Figure 33. The Bolger house in Princeton, Minnesota, has a passive wind system for cooling and ventilation.

Figure 34. Two wind collector kits: **1.** The sailwing by Flanagan's Plans of Astoria, New York, is a downwind horizontal-axis rotor. The generator is housed in the large nose cone. **2.** A downwind system available in plan and kit form from Windworks, Mukwonago, Wisconsin. The generator is in the nose cone, and the tower is an integral part of the system.

is needed. Usually, it is better to recognize your limitations first before entering into a large project. Don't wait until it's too late.

Matching the wind-energy-system components to one another is a crucial factor that determines the overall performance of the system. The wind collector should be appropriate to the energy conversion system and the uses for the energy collected. Similarly, the energy storage system should not be too large or too small for the wind conditions encountered; otherwise, the auxiliary energy system will be used too little or too much. A very large wind energy system may provide so much energy that some might have to be thrown away, while a tiny system may not provide enough energy to be worthwhile.

Some wind energy systems work better in certain areas than others. Windows and ventilators work very well in all wind areas and are the most economical wind energy systems for areas with low wind velocities. Such passive wind systems are relatively straightforward in design and easy to install. Multibladed wind collectors, Savonius rotors, and ducted wind collectors perform well in moderate wind areas. These active wind systems are designed for such conditions and are more complicated and more difficult to install than passive wind systems. Propeller wind collectors and Darrieus rotors perform better in strong wind areas. These active wind systems require high wind velocities to reach the necessary power output levels for economical operation. Such systems are usually the most complex in design, installation, and operation.

Materials are especially important when designing and building a do-it-yourself wind energy system. Most people do not have the necessary tools to machine metals or to make electronic apparatus similar to an off-the-shelf wind energy system. Commonly used materials for do-it-yourself systems are wood, plywood, cloth, cardboard, wire, rope, aluminum, mild steel, bolts, nuts and fasteners from a local hardware store, and materials found in the garage, the basement, the town dump or a lumberyard. Some of the most exciting and workable wind energy systems are developed by people having fun in their own backyards or workshops! Several examples of do-it-yourself projects are shown in Appendix A.

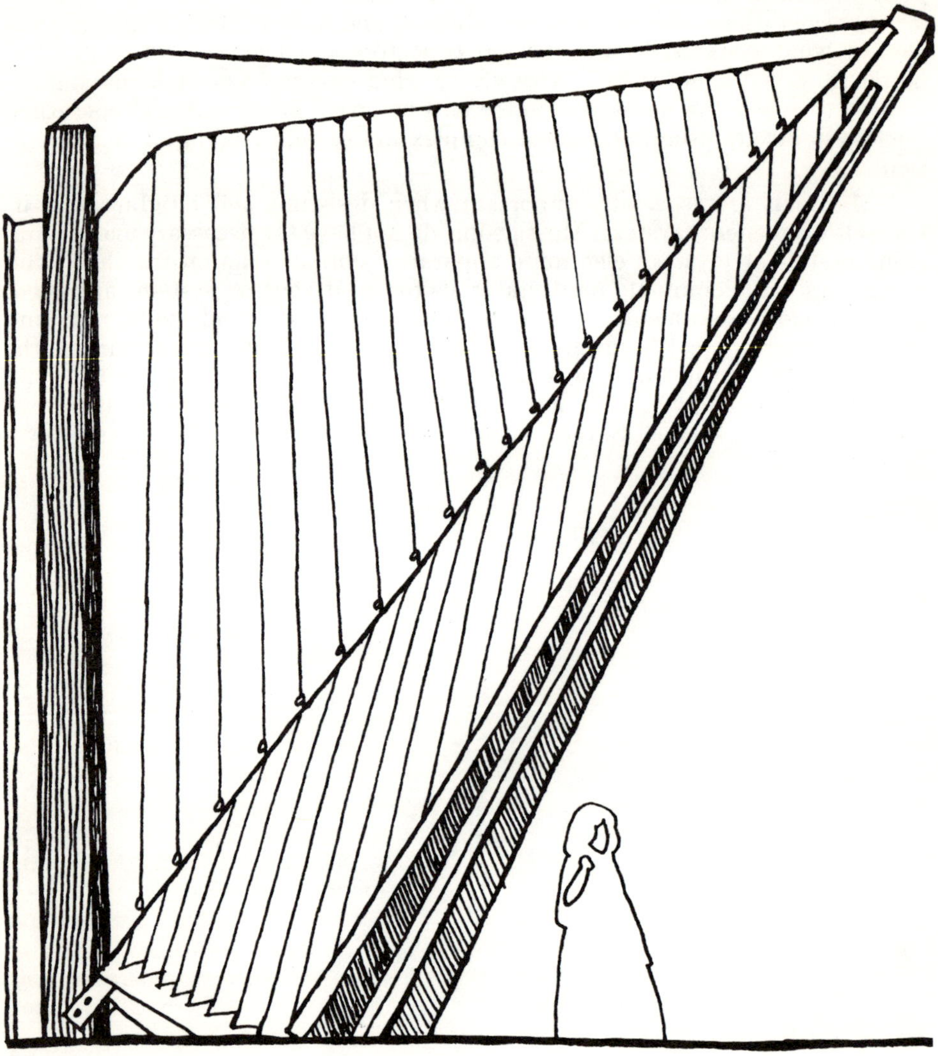

A wind harp uses the wind's energy to make music. This harp was designed and built by Ward McCain and is located in Chelsea, Vermont.

X. Future Developments

Vast amounts of research and development are currently underway for wind energy systems of all types. The present burst of work in the wind energy field has been largely the result of a need for more renewable energy sources, initially to supplement and ultimately to replace dwindling and less renewable fossil fuels. Three critical considerations for the future of wind energy are wind rights, energy storage systems, and wind system manufacturers.

Wind rights are, perhaps, the most touchy consideration. Little or no legal precedent exists pertaining to the use of wind or to someone's right to the use of wind. This right to the use of wind is similar to more widely known and discussed "sun rights." It is reasonable to assume that everyone has an equal right to the use of wind and sun. Yet, it is extremely difficult to prevent one person from blocking or hindering another's access to the wind or sun. There is a general lack of concern for local surroundings whenever a new building or structure is erected. An extreme example of this phenomenon is shown by the continued construction of buildings in front of other buildings in New York City. Some buildings place others in complete shadow. Other buildings reflect blinding sunlight into normally lit spaces, thus changing heating and cooling loads substantially. Buildings create severe winds around themselves and other buildings. People generally assume that builders carefully consider the needs of others when deciding where to locate a new structure. This does not always happen, however, and the necessity for "wind rights" becomes important. How unfortunate it would be, indeed, if a person set up a wind energy system one day only to have it later blocked by someone else's building and rendered all but useless. It is with great anxiety that people now build active wind collector systems. It is not surprising that they attempt to acquire ownership of large areas of land to insure their ability to utilize wind. Current legislation aims at insuring that people wishing to utilize wind energy can do so without interference for as long as they wish.

Energy storage systems are another important consideration for the future of active wind energy systems. Batteries (currently the most commonly used storage system) are expensive, heavy, messy, and sometimes dangerous (due to the hydrogen gas they produce). Expenditures for batteries sometimes account for half of the initial cost of an active wind energy system. The Gemini Synchronous Inverter (made by Windworks) allows excess energy to be fed into the local power grid. This device is not accepted by all utility companies yet, but should be widely accepted and recommended within a few years. The inverter system can do away with the need for an energy storage system entirely, if local electric power is available. This may be very helpful in many cases.

Other energy storage systems currently being developed include hybrid batteries, super flywheels, hydrogen gas, and compressed-air systems. Hybrid batteries should be lighter, less dangerous, longer-lived, and less expensive than today's batteries. Super flywheels store mechanical energy in their rotating discs. The flywheel keeps spinning once it is going; as energy is removed, it slows down. Electrolysis of water produces hydrogen gas, which can be stored and used just as natural gas is used now. Compressed-air storage uses reservoirs or storage tanks to hold compressed air, which is later used in air-turbine generators or is used directly by air-powered equipment.

Wind equipment is manufactured by over 100 firms now. This number will increase as wind use becomes more widespread and acceptable. Large companies, such as those engaged in the manufacture and development of aircraft, are likely candidates to manufacture wind energy systems. Wind systems design and manufacture may become regional as more is learned about the wind characteristics of various areas in the US and throughout the world. Wind system manufacturers will be producing systems in a wide variety of sizes and types so that people will be able to choose the system most suitable to their needs and budget.

Appendices

APPENDIX A: Projects and Experimental Systems

Projects and experiments are fun because they allow us to make things, to watch them work, and to see how the same ideas might be used in other ways. Here are some projects and experiments that are fun and easy to do. They all help to illustrate how wind works and how wind might be utilized to help reduce some of the demands upon earth's limited supply of "conventional" energy sources.

Projects

1. Pinwheel Wind Collector

Materials you will need: a pin, a square piece of construction paper (about 8½ inches by 8½ inches would be fine), and a sharpened pencil with an eraser.

Tools you will need: scissors.

What to do: Lay the square of paper flat on a table and draw a line diagonally from each corner to the opposite corner. Mark the center of the square where the two lines cross and punch a small hole through it with the pencil tip. Next, cut along each line stopping about an inch from the hole in the center of the square. Take the pin and punch a hole in one corner of each of the four flaps—do not punch holes in adjacent corners, but in alternate ones. Pick up a flap at the punched corner and carefully curve it over toward the center

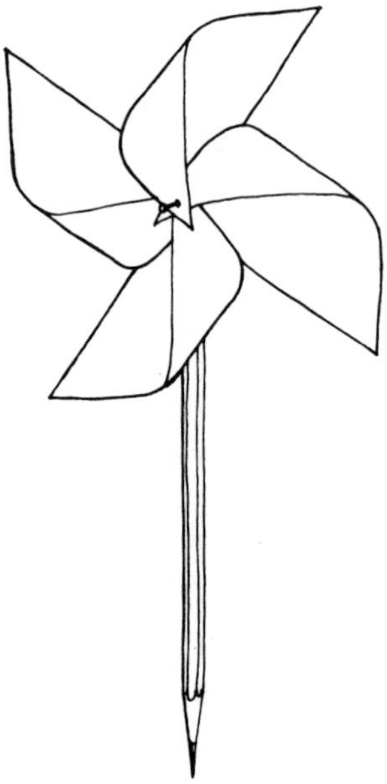

hole. Take the pin and push it downward through the pinhole. Repeat this for the other flaps. When all four flaps are held by the pin, guide the pin with the four flaps on it through the hole in the center of the pinwheel. Hold the pinwheel between your thumb and first two fingers for the next step. Lay the pencil flat on a table and carefully push the point of the pin into the side of the eraser.

Now your pinwheel is complete and ready to go. Pick up the pinwheel near the sharpened tip of the pencil and gently move it through the air. Keep the head of the pin pointed in the direction you are moving the pinwheel. The pinwheel will spin merrily as you move it through the air. You now have a simple wind collector. If the wind were blowing you could hold the pinwheel stationary in the wind and it would spin. The pinwheel is an example of a horizontal-axis active wind collector. It must be pointed into the wind in order to spin.

2. Anemometer

Materials you will need: five 3 oz paper Dixie cups, two straight paper or plastic soda straws, and a straight pin.

Tools you will need: scissors, a paper punch, a sharp pencil with an eraser, and a small stapler.

What to do: Using the paper punch, punch one hole in each of four Dixie cups about 1/2 inch below the rim. Punch four equally spaced holes about 1/4 inch below the rim of the fifth cup. Punch a hole in the center of the bottom of the cup with the four holes. Push a soda straw through the hole in a one-hole

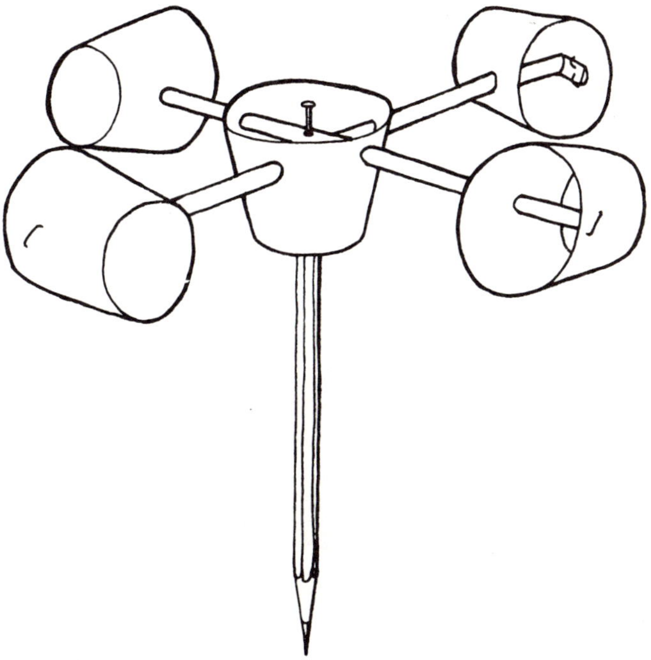

Dixie cup, fold the end of the straw, and staple it to the side of the cup opposite the hole. Repeat this procedure for another one-hole cup and the second straw. Now slide one cup and straw assembly through two opposite holes in the cup with four holes. Push another one-hole cup onto the end of the straw just pushed through the four-hole cup. Bend the straw and staple it to the one-hole cup, making certain that the cup faces in the opposite direction from the first cup. Repeat this procedure using the other cup and straw assembly and the remaining one-hole cup. Align the four cups so that their open ends face in the same direction (clockwise or counterclockwise) around the center cup. Push the straight pin through the two straws where they intersect. Push the eraser end of the pencil through the bottom hole in the center cup. Push the pin into the end of the pencil eraser as far as it will go. Your anemometer is ready to use.

Your anemometer is useful because it rotates at the same speed as the wind. This instrument is quite helpful in accurately determining wind speeds because it gives a direct measure of the speed of the wind. To find the wind speed, determine the number of revolutions per minute of the anemometer. Next calculate the circumference of the circle (in feet) made by the rotating paper cups. Multiply the revolutions per minute by the circumference of the circle (in feet per revolution), and you will have the velocity of the wind in feet per minute. The anemometer is an example of a vertical-axis wind collector. It need not be pointed into the wind to spin.

3. Wind Direction Indicator

Materials you will need: one straight paper or plastic soda straw, one piece of construction paper, a pencil with an eraser tip, and one straight pin.

Tools you will need: a stapler and scissors.

What to do: Cut one end off the piece of construction paper so that it is square. Fold one corner of the square over until it meets the opposite corner to form a large triangle. Crease the fold and open the paper. Cut along the fold to

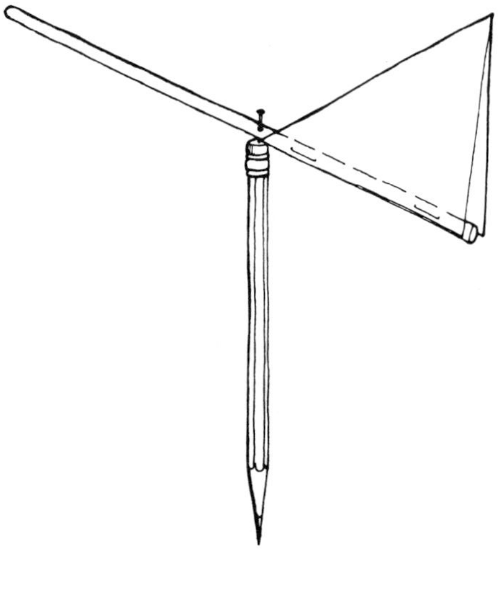

make two triangles. Fold one triangle in half once again and crease it along the fold. Now place an open edge of this folded triangle over the soda straw with the point toward the center of the straw and the other open edge at the end of the straw. Staple the tail to the straw. Next push the pin through the soda straw about one inch ahead of the front of the tail. Push the pin into the top of the eraser on the pencil. Your wind direction indicator is now ready to go.

Hold the wind direction indicator in the wind. It automatically turns around until the tail of the straw points away from the wind and the tip points into the wind. This instrument is useful in determining where the wind is coming from at any time and in noting variations during the day or from season to season. Hold your wind direction indicator in the wind and notice how often the wind direction changes.

4. Wind Collector System Chart

Materials you will need: pencil, pen, felt-tip marker, one piece of poster board, and several sheets of writing paper.

Tools you will need: ruler and straight edge.

What to do: Make a list of all the different kinds of wind collector system you can think of. Divide the list into *passive* and *active* systems. Make a chart showing how each collector fulfills the needs for wind energy systems. Remember that each system includes collector, support, orienting device, control, convertor, distribution system, storage, use, auxiliary systems, maintenance and safety provisions. What other systems would be practical to design and build? What kinds of wind energy systems are most practical for your area?

Experimental Wind Systems

Experimental wind systems may be fashioned from recycled materials and pieces of junk. A local junkyard or machine shop may yield such valuable items as windows, fan blades, bearings, shafts, pulleys, towers, generators, pumps, sheet metal, and batteries. These pieces can be combined to produce experimental wind energy systems.

When using recycled materials, be sure that all parts are properly cleaned and reconditioned. This will ensure the best performance and safety of the experimental wind system. Active wind systems require special care. They produce large amounts of electrical or mechanical energy that may be harmful if not properly controlled. Wood is one of the most useful materials for experimental wind systems. Wood can easily be shaped and refitted, allowing for quick and easy changes in the system.

1. Windows

One of the easiest passive wind collector systems to experiment with is a window. Two pieces of window sash and a wood frame wall to mount them in provide the basis for many types of window experiments. Hardware such as hinges, screws, window latches, and nails, and tools such as a hammer and a screwdriver, are also helpful. Several types of windows to experiment with are single-hung, double-hung, single-casement, double-casement, single-awning, double-awning, high- and low-awning, hopper, vertical-pivot and horizontal-pivot windows. Each window type has special wind characteristics well suited to different uses. A horizontal pivot window is shown in Figure 10.

2. Wind Electric System

A wind electric system requires a propellerlike rotor with two or three narrow blades. Attach this rotor to the shaft of a permanent-magnet generator. Lubricate the generator bearings until the rotor spins freely. Attach about ten feet of wire to each contact of the generator. Mount this assembly atop a post or short tower so that the blade tips of the rotor clear the ground by at least five feet. Now connect a knife switch to one lead from the generator. Open the knife switch and secure it. Connect the other lead from the generator to one contact of a light socket. Connect one end of a short piece of wire to the remaining contact of the knife switch and the other end to the remaining light socket contact. Screw a 15- or 25-watt bulb into the light socket. Turn the wind collector into the wind and close the knife switch. If the wind is blowing with sufficient velocity, the small bulb will light. The stronger the wind, the brighter the bulb. The wind collector can be made to turn into the wind automatically by adding a tail and mounting a bearing between the generator and the tower or post.

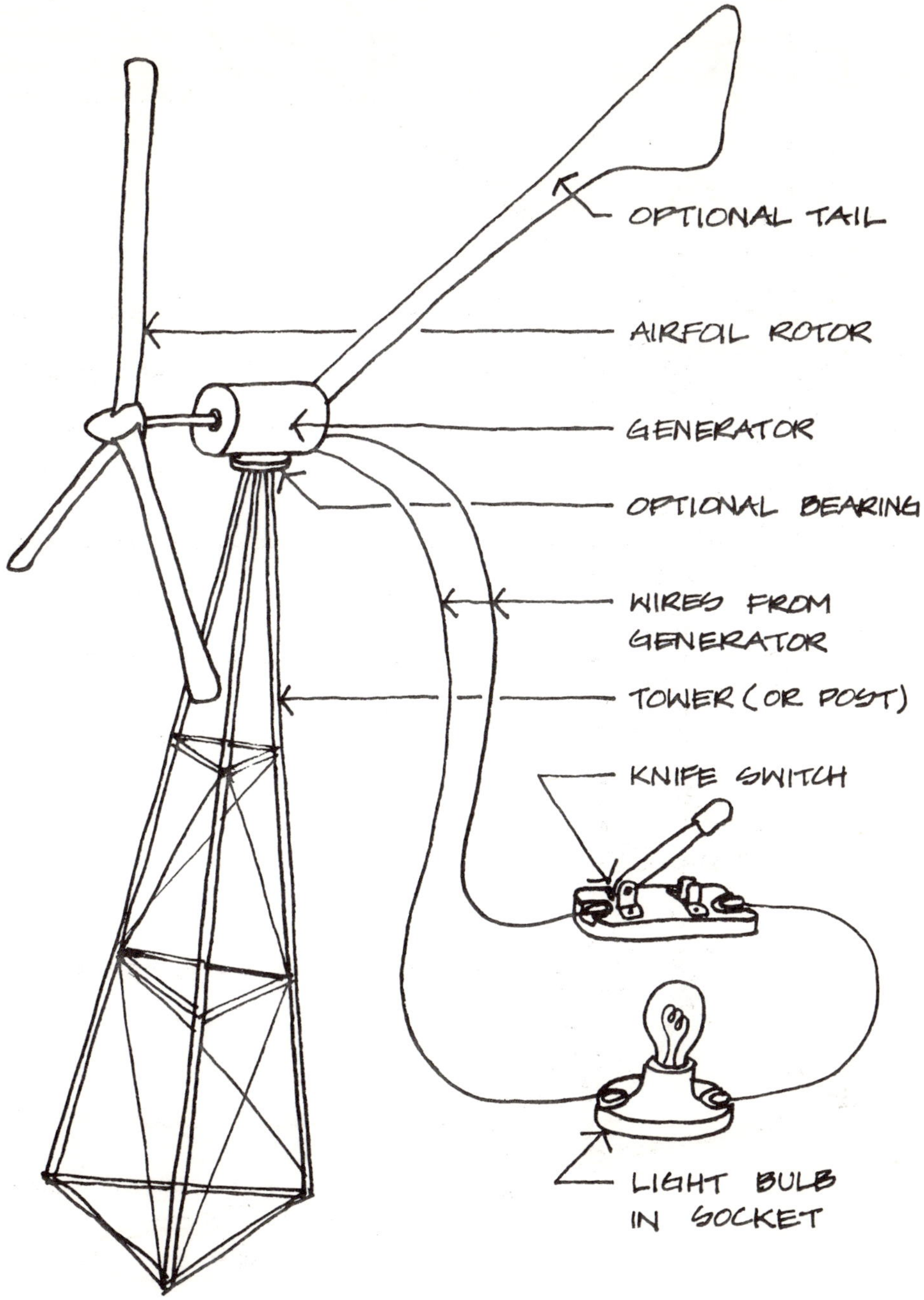

WIND-POWERED ELECTRIC SYSTEM

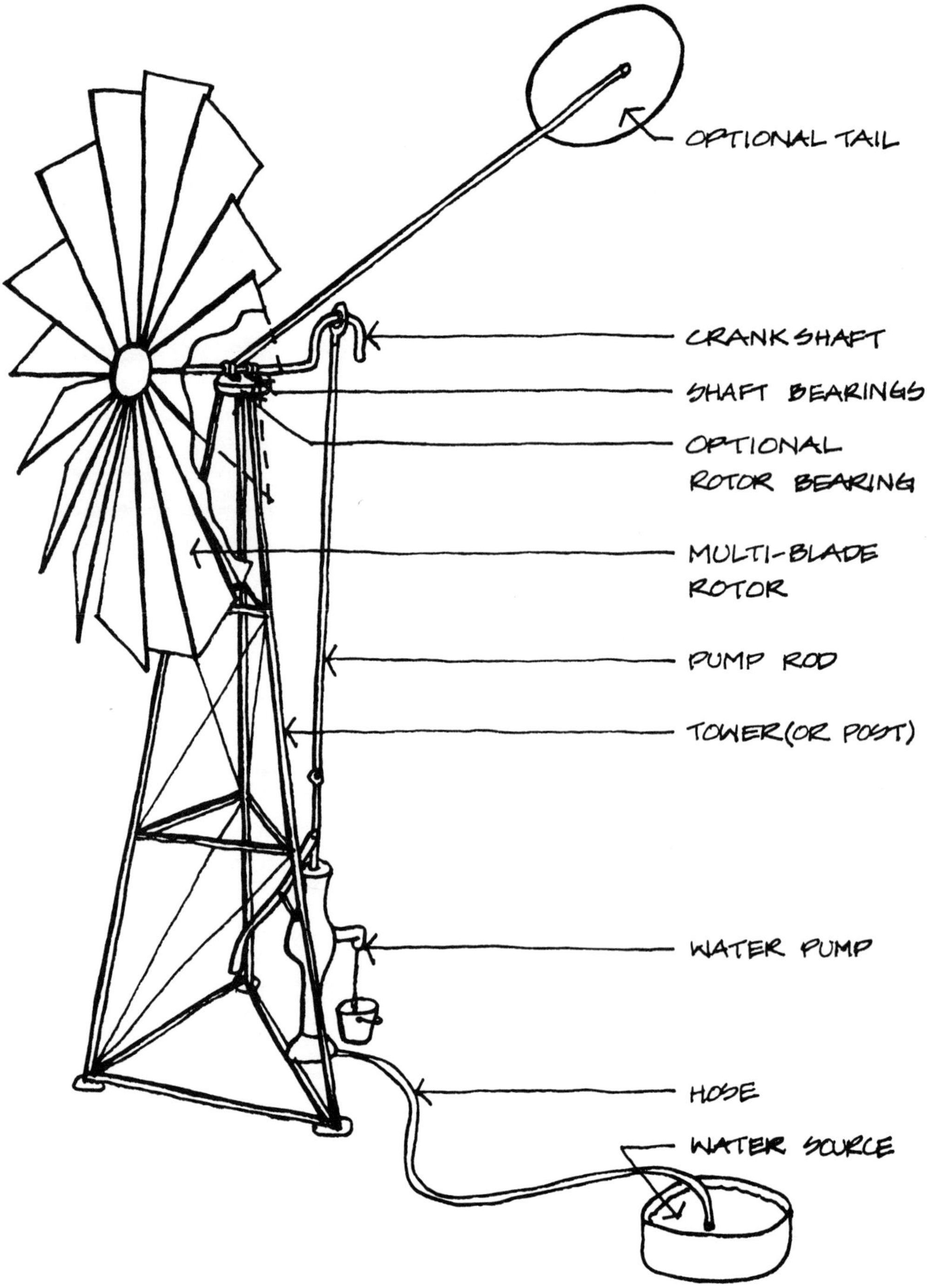

WIND-POWERED WATER PUMP

3. Wind-Powered Water Pump

A wind-powered water pump requires a propeller rotor with either ten or twelve wide blades of metal or wood. Attach this rotor to a crankshaft with two bearings on it. Slip an eight-foot rod over the crank. Mount the bearings and rotor on the top of a post or tower so that the blade tips of the rotor clear the ground by at least five feet. Secure a small hand-operated water pump to the ground. Attach the crank rod to the water-pump lever and give the wind rotor a spin. Lubricate the pump bearing until it moves freely and easily. Attach a hose to the pump inlet and place the other end of the hose into a small tank of water. Now face the wind rotor into the wind and give it a small push. If the wind is strong enough, the rotor will continue to spin. It will soon begin to pump the water from the tank. The stronger the wind, the faster the water will be pumped.

APPENDIX B: Passive Wind Design

1. Carefully planted trees and other vegetation can substantially reduce heat loss due to airflow. The equation given here describes the interrelationship between heating load, the difference between inside and outside temperatures, distance to the shelterbelt, and wind velocity:

$$Q = 1.3\,(10^{0.018 \times L \times 0.07 \times V}) \times \Delta T$$

Q is the heating load in BTU per hour (BTUh), L is the ratio of the distance of the barrier to its height, V is the wind velocity in mph, and ΔT is the temperature differential.

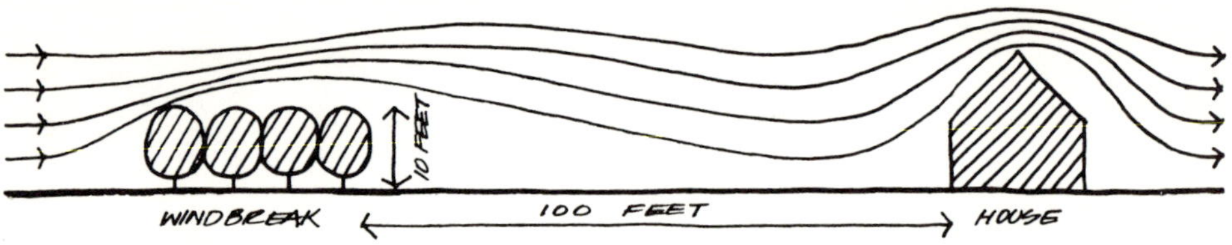

As an example, let's assume a 50°F temperature differential, a 10 mph wind, and a 10 ft high shelterbelt 100 ft away (L = 100/10 = 10). Then,

$$\begin{aligned} Q &= 1.3\,(10^{0.018 \times 10 \times 0.07 \times 10}) \times 50 \\ &= 1.3\,(10^{.18 \times .7}) \times 50 \\ &= 1.3\,(10^{.126}) \times 50 \\ &= 1.3 \times 1.34 \times 50 \\ &= 87 \text{ BTUh heat loss} \end{aligned}$$

If the shelterbelt were only 10 ft away from the building, the heat loss would be 67 BTUh, a 23% reduction.

2. The airflow through an inlet such as a window depends on the wind velocity V and the ratio of the air outlet to the inlet:

$$R = 3150\,A \times V$$

where R is the rate of airflow in cubic feet per hour, A is the inlet area in square feet, and V is the wind velocity in mph. In this equation we assume that the wind is perpendicular to the inlet and that the inlet and outlet areas are equal. As an example, let us calculate the airflow in an 800 ft^3 room with a 5 ft^2 window and a wind speed of 5 mph.

$$\begin{aligned} R &= 3150 \times 5 \times 5 \\ &= 78{,}750\ ft^3/hr \\ &= 1312\ ft^3/min \end{aligned}$$

Thus, the room's air is changed completely about one and one half times per minute.

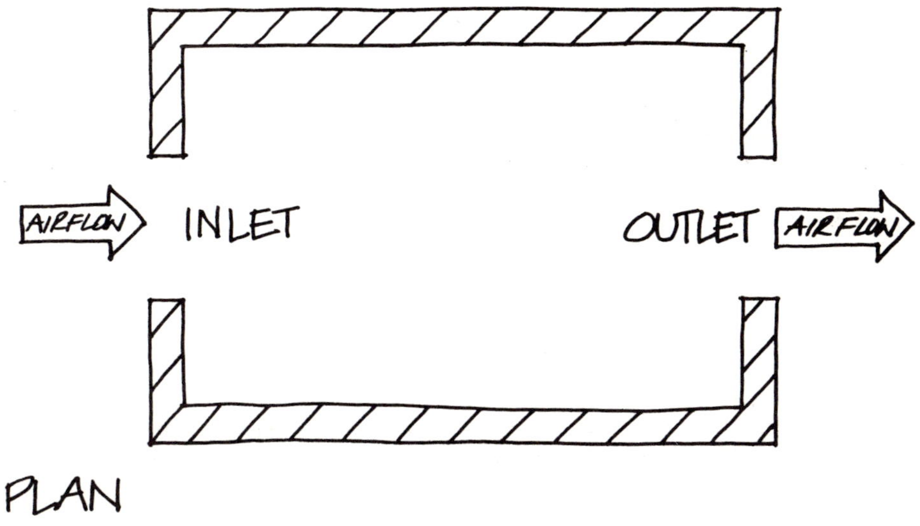

If the inlet and outlet areas are not the same, the constant in the equation changes:

Ratio of outlet to inlet area	Constant
1:1	3150
2:1	4000
3:1	4250
4:1	4350
5:1	4400
3:4	2700
1:2	2000
1:4	1100

Accordingly, the greater the ratio of outlet area to inlet area, the greater the airflow volume. Outlet area is very important for adequate air circulation.

3. In Section 2, we assumed that the window was perpendicular to the direction of the wind. This is the most effective orientation; diagonal orientations are much less efficient. The efficiency of a particular orientation can be approximately determined from the figure below:

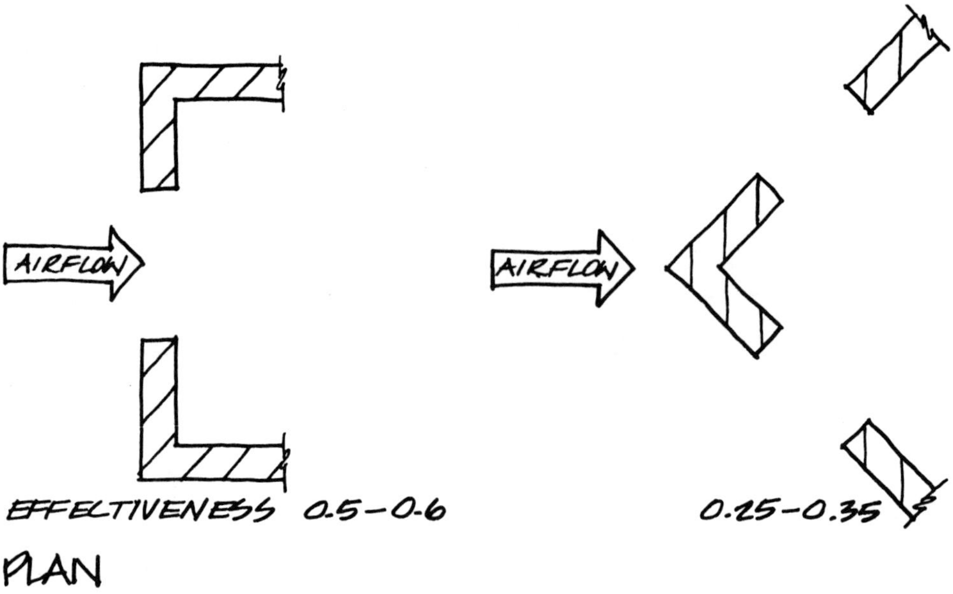

For our example,

$$R = 3150\,E \times A \times V$$

where E is the effectiveness of the openings. Let's assume that the efficiency of the perpendicular orientation of the example in Section 2 is .55 and of a diagonal orientation is .30. Then, the actual* flow through the room with an inlet presented perpendicular to the wind is

$$\begin{aligned} R &= 3150 \times .55 \times 5 \times 5 \\ &= 43{,}312\ \text{ft}^3/\text{hr} \\ &= 722\ \text{ft}^3/\text{min} \end{aligned}$$

The flow through the room with the diagonally presented window is

$$\begin{aligned} R &= 3150 \times .30 \times 5 \times 5 \\ &= 23{,}625\ \text{ft}^3/\text{hr} \\ &= 394\ \text{ft}^3/\text{min} \end{aligned}$$

The airflow is almost twice as great when the inlet is perpendicular to the direction of the wind. Note that the two diagonal windows are each half the size of the

*Notice that this is less than the value we got in Section 2 where we assume perfect efficiency.

single perpendicular window. If the two windows in the diagonal orientation are both the same size as the perpendicular window, the airflow is twice as great or 788 ft^3/min.

4. Inlet placement governs the basic flow pattern within the ventilated space. A high inlet gives greatest flow near the ceiling; low, near the floor; and middle, through the center of the room. Outlets may be high, central, or low so long as inlets are located for the desired airflow, usually central or near the floor.

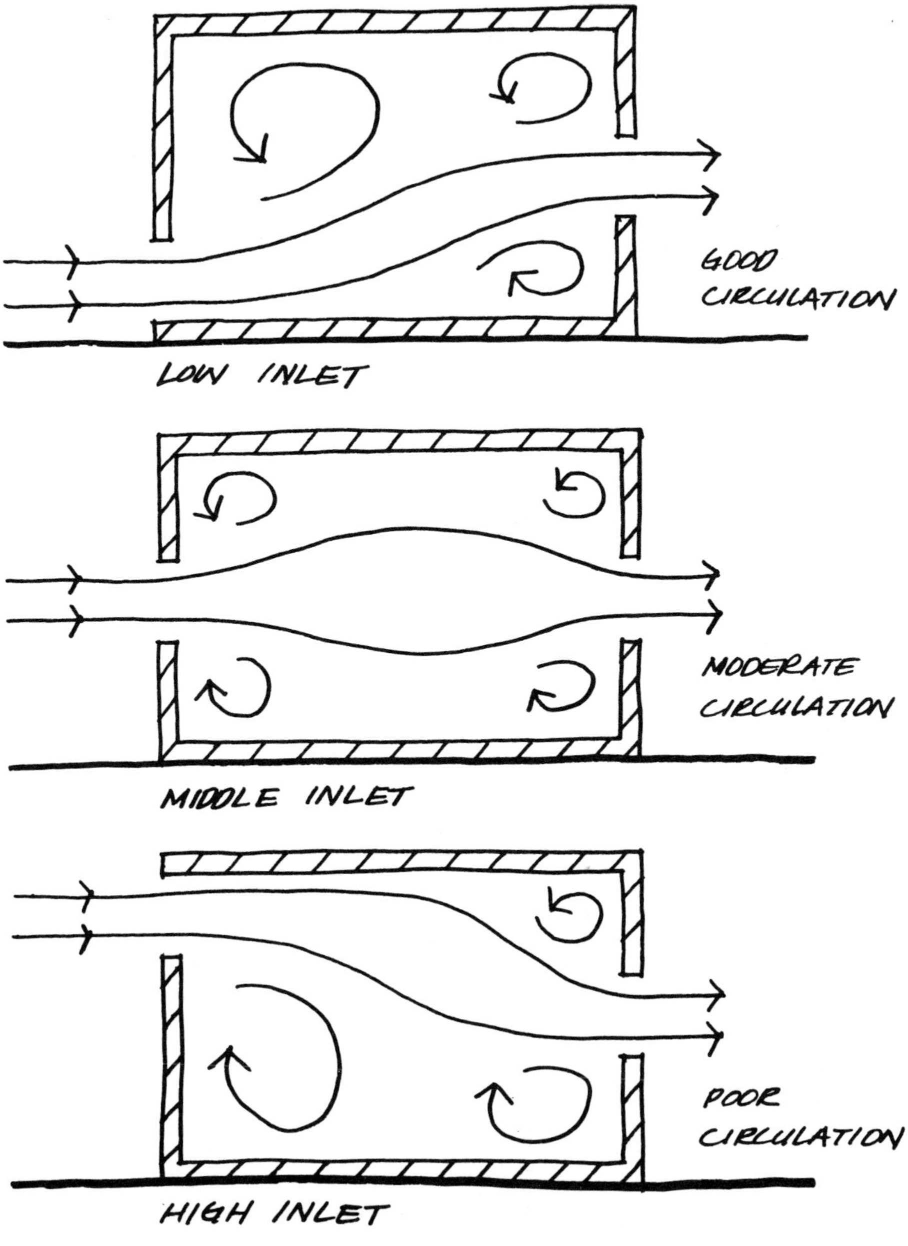

5. Overhangs and deflectors on structures enhance airflow by collecting air that would otherwise escape. Trees, bushes, fences, and walls may all serve as deflectors; they act to scoop the moving air into the inlet before it can go around the structure.

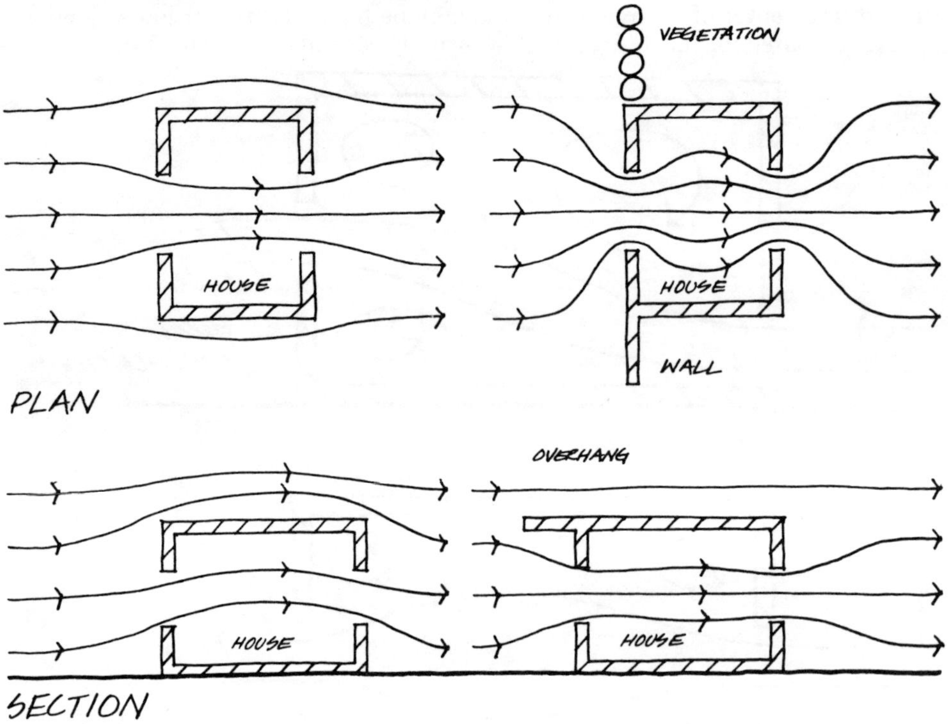

6. Warm air rises. We can use this phenomenon to ventilate a room without using electric fans. Ventilation by temperature differential also depends on the ratio of outlet area to inlet area and the difference in height between outlet and inlet:

$$R = 540 \times A \times \sqrt{H\,(t_i - t_o)}$$

where R is the rate of airflow, A is the inlet area, H is the difference in inlet and outlet height, and t_i and t_o are the inside and outside temperatures respectively. Let's assume we have a room with two windows whose areas are both 5 ft^2; that the outlet window is four feet above the inlet; and that the indoor temperature is 75°F and the outside temperature is 70°F. Then,

$$\begin{aligned} R &= 540 \times 5 \times \sqrt{4\,(75° - 70°)} \\ &= 540 \times 5 \times \sqrt{20} \\ &= 12{,}074 \text{ ft}^3/\text{hr} \\ &= 201 \text{ ft}^3/\text{min induced ventilation.} \end{aligned}$$

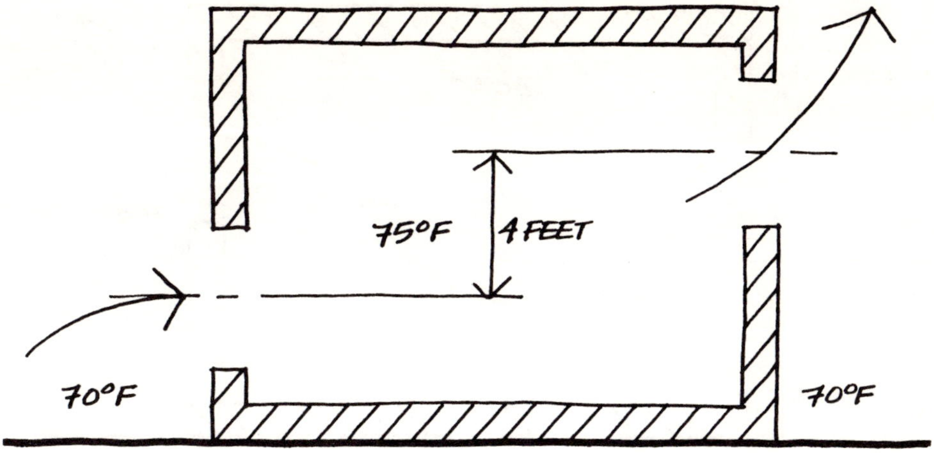

As with the example in Section 2, the constant in this equation changes according to the ratio of outlet area to inlet area:

Ratio of outlet to inlet area	Constant
1:1	540
2:1	680
3:1	720
4:1	740
5:1	745
3:4	455
1:2	340
1:4	185

7. This induced circulation produces a slight wind that has a cooling effect. We can compute the velocity of this wind for our above example according to the following equation:

$$V_i = \frac{R}{P} \times H$$

where V_i is the velocity of the induced wind, R is the flow rate (201 ft^3/min), P is the volume of the airflow pattern (800 ft^3), and H is the difference in inlet and outlet height. From our example,

$$\begin{aligned} V_m &= \frac{201}{800} \times 4 \\ &= 1 \text{ ft/min.} \end{aligned}$$

8. The temperature differential within a room can also stimulate air movement by the chimney effect. Warm air leaves the room by the high outlet, and cooler air is sucked into the low inlet. The flow rate is

$$R = 9.4 \times A \times \sqrt{H \times \Delta T}$$

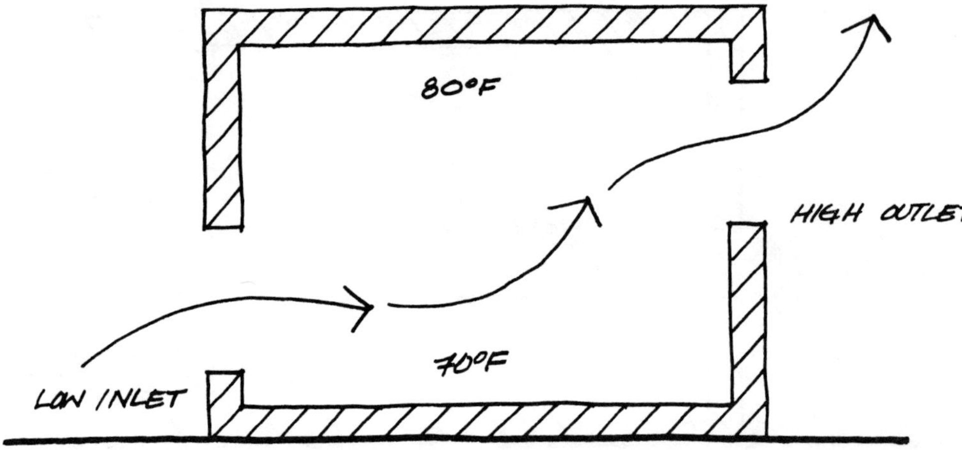

where ΔT is the difference in temperature between floor and ceiling. Let us assume that this difference is 10°F and that we are using the same room as before. Then,

$$R = 9.4 \times 5 \times \sqrt{4 \times 10}$$
$$= 297 \text{ ft}^3/\text{min}$$

APPENDIX C: Active Wind Design

1. There are many different kinds of active wind collectors, and a considerable number have been illustrated in this book. These collectors can be classified according to a variety of characteristics.

Probably the most basic distinction is whether the collector is activated by lift or drag. *Drag* is the pushing force of the wind; it's the force you feel when the wind is blowing against you. The anemometer you built in Project 2 is an active collector that operates by drag, and so is the Savonius rotor illustrated at the beginning of Chapter IX. Such devices are cheap and easy to manufacture, but they are not so efficient as lift devices.

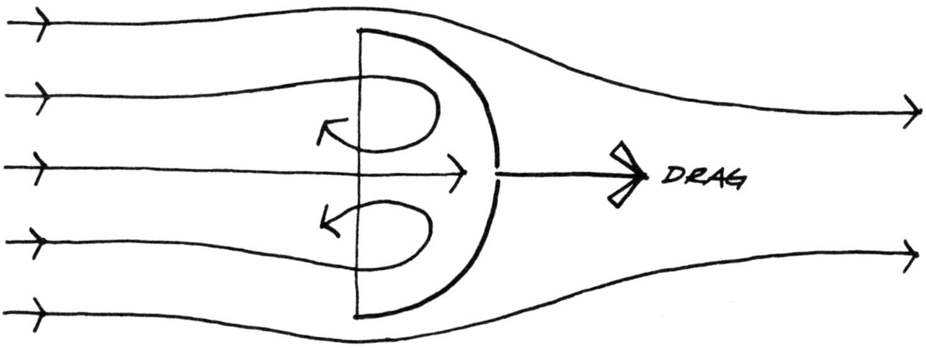

You have noticed how a wind can raise leaves and paper off the ground. This force is the *lift*. Many active wind collectors use the lift of the wind to operate. These machines have blades that resemble airplane wings—that is, they use airfoils. The air flowing past an airfoil creates both lift and drag forces. In a propeller-driven airplane, the airflow past the wings is supplied by the propeller blades (which are themselves airfoils), and the propellers pull the airplane through the air. This airflow creates the lift force on the wings that keeps the plane aloft. In the case of the airplane and most lift wind collectors, the drag force is a nuisance that must be reduced as much as possible. The Darrieus-Savonius rotor in Chapter III uses both drag and lift devices: The Savonius is a drag device, and the Darrieus is a rather odd-looking lift device.

There are other distinguishing characteristics. A wind machine may have a horizontal axis (like most windmills), a vertical axis (the Darrieus and Savonius rotors), or an inclined axis (Figure 13). The rotors may face upwind (Figure 32), downwind (Figure 15), or crosswind (Figure 11). They may be free-flow collectors or concentrating collectors like the venturi ducted wind turbine shown at the beginning of Chapter IV. The rotors may have one, two, three, or four blades; they may be multibladed such as the "bicycle-wheel" rotor (Figure 17). The

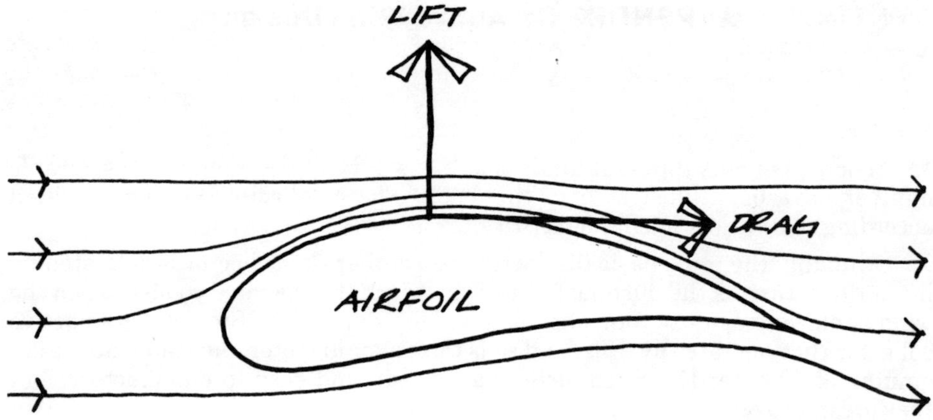

collector itself may be fixed (unidirectional) or it may be a tracking or orientable machine.

2. The type of wind collector to use depends on the average wind velocity at the site:

Average wind velocity in mph	Collector type
<4	Passive and possibly concentrating
4-10	Passive and concentrating
>10	Passive, concentrating, and free flow

3. We saw in Chapter IV that the power, and hence, the energy available in the wind are determined by the air density ρ, the collector area A, and the velocity of the wind cubed V^3. Since the air density doesn't vary very much, we can include it in the constant in the following equation for the theoretical maximum power in the wind:

$$P_{Tmax} = 1.68 \times 10^{-6} A \times V^3 \quad \text{(for area in ft}^2\text{ and velocity in ft/sec)}$$
$$= 5.3 \times 10^{-6} A \times V^3 \quad \text{(for area in ft}^2\text{ and velocity in mph)}$$

For instance, the maximum power in a 20 mph wind for a collector of 100 ft^2 is

$$P_{Tmax} = 5.3 \times 10^{-6} \times 100 \times 20^3$$
$$= 5.3 \times 10^{-4} \times 8 \times 10^3$$
$$= 4.24 \text{ kW}$$

4. No machine is perfect, and even the ideal machine cannot extract all the energy available in the wind. The maximum available power is only about 60 percent of the theoretical maximum power in the wind:

$$P_{Amax} = .59\, P_{Tmax}$$
$$= 1 \times 10^{-6} A \times V^3 \text{ (for A in ft}^3\text{ and V in ft/sec)}$$
$$= 3.16 \times 10^{-6} A \times V^3 \text{ (for A in ft}^3\text{ and V in mph)}$$

In real machines even less power is actually extracted, and there are additional losses. For instance, at times the wind is so light that the rotor cannot overcome its own inertia and frictional forces; sometimes the wind is so strong that the machine must be stopped to prevent damage. Even under the conditions the machine is designed for, there are power losses from drag, eddies, friction, and so on. (See Chapter V, especially Figure 18.) Each time one form of energy is converted to another there are power losses. Each piece of equipment has its own efficiency. The actual output at the generator is

$$P_{actual} = E_{rotor} \times E_{drive} \times E_{generator} \times P_{Amax}$$

where the Es are the efficiencies of the machine's components (Figure 18).

5. Wind data include the wind speed, its direction, and the amount of time (expressed as a frequency or percentage) the wind has a particular speed and direction. The maximum extractable energy available from the wind, with speed i and direction j, per square foot of the collector is

$$\frac{E_{max}}{ft^2} = \frac{P_{ij}}{ft^2} \times T_{ij}$$

where P_{ij} is the power at wind speed i and direction j and T_{ij} is the time the wind has speed i and direction j. From our example, we saw that a 20 mph wind has a maximum power of 4.24 kW at an area of 100 ft^2, or $\frac{4240}{100} = 42.4\ W/ft^2$. Suppose that the wind comes from the north-northeast at 20 mph three percent of the time. Then, since there are 8760 hours per year, we can expect

$$\frac{E_{max}}{ft^2} = 42.4 \times .03 \times 8760 = 11.1\ kWh/yr$$

for winds with this speed and direction.

6. To find out how much energy there is at all wind speeds and directions we must add the product P_{ij}/ft^2 and T_{ij} for all values of wind speed i and direction j. This is indicated by the following equation:

$$\frac{E_{max}}{ft^2} = \sum_{ij} \frac{P_{ij}}{ft^2} \times T_{ij}$$

where the symbol $\sum_{ij}$ means that we are summing the products of P and T for all wind speeds and directions. This can be quite a tedious procedure, but it may be necessary if we are building a unidirectional wind collector, since we would want to orient it in the direction that permits the greatest wind power extraction.

If the collector can orient itself into the wind, the problem is easier. The energy available from all directions per square foot of collector is

$$\frac{E_{max}}{ft^2} = \sum_{i} \frac{P_{ij}}{ft^2} \times T_{i}$$

Here, we need to sum only over the velocity.

7. We can further simplify the problem of predicting how much energy we can extract. Rather than sum for *every* velocity, we can use average velocities for a number of velocity intervals. That is, we can use an average for velocities in the intervals 0-3 mph, 3-6 mph, 6-9 mph, and so on, instead of computing for each velocity. The average velocity used to calculate P over each velocity range is

$$V_{avg\ P} = \langle V \rangle = \left[\frac{V^3{}_{min} + V^3{}_{max}}{2}\right]^{1/3}$$

We define the average this way since the power varies as V^3 and the usual average $(V_{min} + V_{max})/2$ would underestimate the power available in each velocity range. Let's say we want to compute $\langle V \rangle$ for the velocity range 8-12 mph. Then

$$\begin{aligned} \langle V \rangle &= \left(\frac{8^3 + 12^3}{2}\right)^{1/3} \\ &= \left(\frac{512 + 1728}{2}\right)^{1/3} = \left(\frac{2240}{2}\right)^{1/3} = (1120)^{1/3} \\ &= 10.4 \text{ mph} \end{aligned}$$

Complete wind data for a site are necessary to make the above calculations. Such information is available for a variety of weather station sites and is usually published in chart form. The chart lists the percentage of the time the wind has a given direction and velocity range on a monthly or yearly basis.

8. At first, one might expect that the most energy will be extracted from the average wind. However, since the power available (and thus the energy) depends on the cube of the velocity, the greatest energy extraction occurs at about 1.7 V_{avg} for an average wind site. (See figure below and Figure 6.) This fact is im-

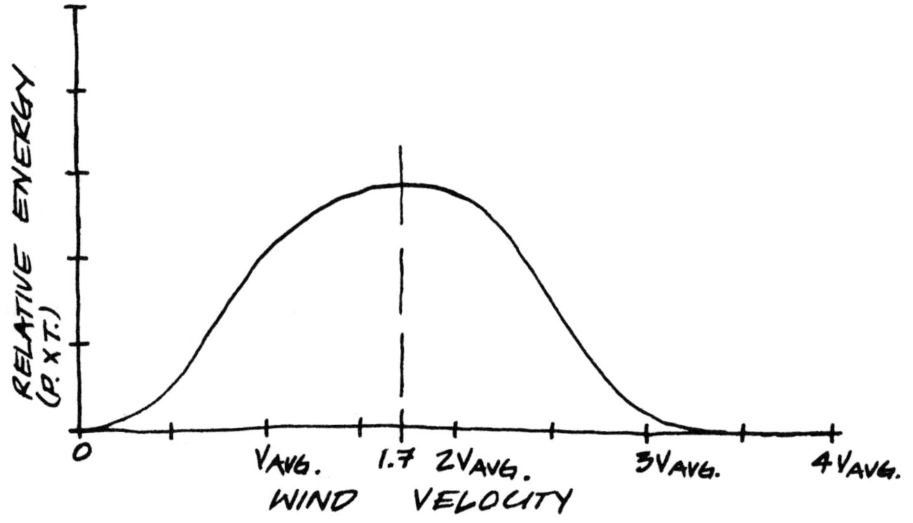

portant in designing a wind energy conversion system, since one should design for maximum efficiency at 1.7 V_{avg} or the velocity of greatest available energy.

9. The actual power output from a wind collector system can be determined from the power output vs wind velocity curve for the wind collector. These curves are published for many wind systems, and they may be determined experimentally in a wind tunnel or at the site.

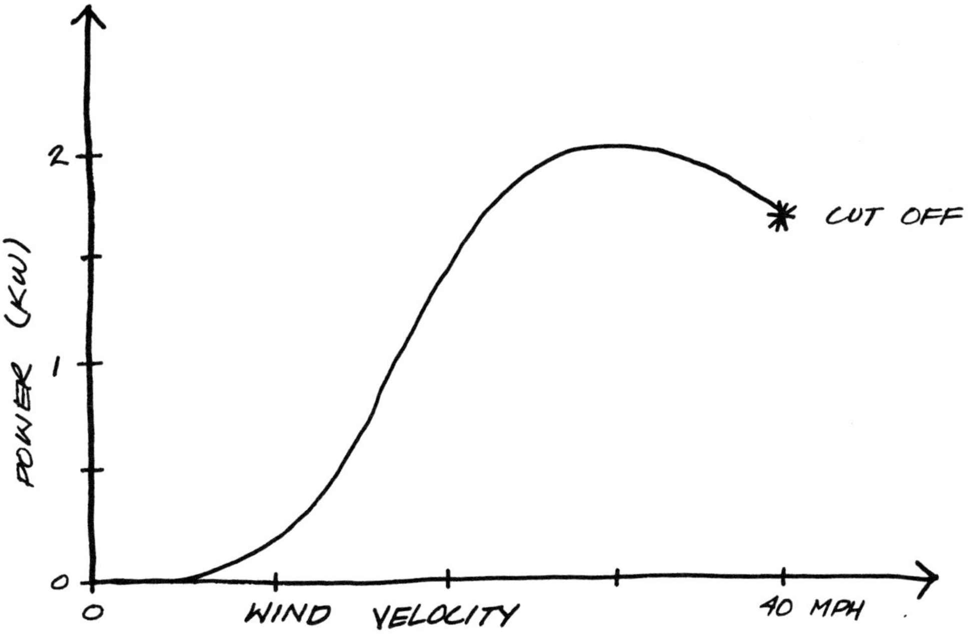

10. The actual energy output from the wind system may be obtained from the power distribution curve in Section 9 and from the velocity distribution curve for the site. The time is (as in Section 5) the frequency that the wind speed is in a certain interval; this information is read from the wind data chart. Then,

$$E_{actual} = \sum_i P_i \times T_i$$

where P_i is the actual power at velocity i (as read from the power vs velocity curve) and T_i is the frequency of occurrence of velocity i (from the wind chart).

11. If there is no frequency of occurrence chart for your site, but only an average wind velocity, you may estimate the energy output by using the frequency of occurrence chart for a different site with the same average velocity as your site. (See Figure 6.)

12. A detailed monthly analysis of a wind collector system for your site can be done with the aid of the following wind collector calculation chart. Fill in the chart with the wind velocity information for your site and proceed from A through P. The completed chart will give the useful energy expected from your wind system. An example is shown for the month of January; the average wind velocity is taken to be 10 mph, the machine is a 16 ft diameter propeller collector, and there are 360 hours during the month when the machine is actually collecting energy.

Wind Collector Calculation Chart

Site elevation (ft)	0	1000	2000	3000	4000	5000
Mass density of air ρ (slug/ft^3)	.00238	.00231	.00224	.00218	.00211	.00204
Constant $K = \rho \times 7.06 \times 10^{-4} = 1.6 \times 10^{-6}$						

1. From the list at left, find the mass density of the air at your site's elevation.
2. Multiply that value by the conversion constant 7.06 $\times 10^{-4}$ to find constant K.
3. Use that value for K in the calculations below.

Column	A	B	C	D	E	F	G	H
Month	Average wind velocity V (ft/sec)	V^3 (A × A × A)	KV^3 (kW/ft^2)	Collector efficiency (decimal)	Converter efficiency (decimal)	Distribution efficiency (decimal)	System efficiency (D × E × F)	Collection time (hr/mo)
Jan	14.67 (10 mph)	3154	.005	.45	.75	.75	0.25	360
Feb								
Mar								
Apr								
May								
Jun								
Jul								
Aug								
Sep								
Oct								
Nov								
Dec								

Column	I	J	K	L	M	N	O	P
Month	System output ($C \times G \times H$)	Usable energy collected ($2 \times I$)* (kWh/ft²/mo)	Wind collector area (ft²)	Wind energy available ($J \times K$) (kWh/mo)	Energy required (kWh/mo)	Auxiliary energy required ($M - L$) (kWh/mo)	Wind energy used (kWh/mo)	Use factor ($O \div M$) (percent)
Jan	0.45	0.9	201	182	300	118	182	60
Feb								
Mar								
Apr								
May								
Jun								
Jul								
Aug								
Sep								
Oct								
Nov								
Dec								
Totals								

*Note that this chart doubles the actual energy output of the system since the *average* wind velocity is used in column J.

Information sources: local weather bureau, *Weather Atlas of the United States*, *Standard Handbook for Mechanical Engineers*, wind velocity distribution records, estimated efficiencies of system components, utility records, and energy consumption bills.

APPENDIX D: History of Wind Use

Wind has been available for a long, long time. It has been used for a myriad of things ranging from ventilation and transportation to the generation of electricity. A chronology of the highlights in the development of wind energy systems follows. There is a great diversity in the systems, and current research and developments are so wide ranging that a complete chronological listing would be quite confusing and difficult to compile.

A CHRONOLOGICAL SUMMARY OF THE USE OF THE WIND

DATE	USE	LOCATION	AXIS	SUPPORT	COMMENTS
Long Ago	Ventilation Cooling Transport	World (as we'll never know it!)	—	Shelter Sailboat	—
644 AD	Grinding	Persia	V	Building	Open turbine
915	Grinding	Persia	V	Building	Ducted turbine
1000	Irrigation	China	V	Free-standing	Feathering blades
1000	Grinding	Crimea	V	Building	Feathering blades
1180	Grinding	France	H	Post	Four sail blades covered with cloth
1187	Grinding	England	H	Post	” ”
1200	Grinding	Europe, Australia	H	Post	” ”
1250	Grinding	Azores, Mediterranean area	H	Building	Triangular cloth sails on spars; 8 to 12 sails
1420	Grinding	England, France	H	Building	” ”
1430	Drainage	Holland	I	Open	Water pump
1450	Drainage	Holland	H	Building	Water pump
1600	Drainage	Europe, Africa	H	Building	Water pump
1750	Sawing, oil pressing, papermaking, hammering	Holland	H	Building	Mechanical devices
1772		Scotland	Spring sail, shutter-adjusting mechanism, invented by Andrew Meikle		
1775		England	Fantail, automatic orientation device, invented by Edmond Lee		

KEY: H = Horizontal; V = Vertical; I = Inclined

1787		England	Centrifugal governing device introduced by Thomas Mead		
1789		England	Roller blind, remote-control sails, invented by Stephen Hooper		
1807		England	Patent sail, remote-control shutters on blades, invented by Sir William Cubitt		
1854	Pump	US	H	Tower	Annular sails
1860	Pump	England	Air brakes applied to patent sails by R. Catchpole		
1890	Electric	Denmark	H	Tower	Patent sails and twin fantails
1900	Electric	Holland	H	Building	Adapted corn mills
1920		France	Three-blade airfoil experiments by G. Darrieus		
1927		Netherlands	Application of airfoils on leading edges of blades by A.J. Dekker		
1931	Electric	Crimea	H	Tower	Three-blade airfoil
1933		Finland	V	Savonius rotor invented by S.J. Savonius	
1940	Electric	US	H	Tower	Grandpa's Knob, near Rutland, Vermont
1950	Electric	England, France, Russia	H	Tower	Three-blade airfoil
1955	Electric	Denmark	H	Tower	Three-blade airfoil with bowsprit guy support by J. Juul

—LULL IN WIND-USE RESEARCH—

1966	Electric	US	H	Tower	Two- or three-blade sailwing wind collector by Prof. Sweeney
1968	Electric	US	H	Tower	Multivane bicycle-wheel wind collector
1970	Electric/ mechanical work	USSR	H & V	Tower, earth, building	Ducted collectors increasing wind velocity 1.5 to 6.0 times
1974	Electric/ heating	US	H	Tower, building	Venturi ducted wind dynamo by Bill Gillette
1976	Electric	US	H	Tower	Large (125 ft dia.) two-bladed wind turbine built near Sandusky, Ohio

FOR FURTHER READING

There are hundreds of books, pamphlets, and articles on wind energy. Here is a listing of some of the more pertinent and useful ones.

Allison, H.J., et al. *Energy Storage Research at Oklahoma State University.* Proceedings of the Midwest Power Symposium at Iowa State University, Ames, Iowa, October 1970.

Aynsley, R.M. "Wind Effects Around Buildings." *Architectural Science Review,* March 1972.

Aynsley, R.M. "Wind Effects on High- and Low-Rise Housing." *Architectural Science Review,* September 1973.

Burke, Barbara, and Robert N. Meroney. *Energy from the Wind: An Annotated Bibliography.* Fort Collins, CO: Colorado State University, 1976.

Cermak, J.E., et al. *Problems of Atmospheric Shear Flows and Their Laboratory Simulation.* Fort Collins, CO: Colorado State University, 1966.

Clegg, Henry. *New Low-Cost Sources of Energy for the Home.* Charlotte, VT: Garden Way Publishing, 1975.

Clews, Henry. *Electric Power from the Wind.* Solar Wind Publications, 1973.

Coonley, Douglas R. *Design with Wind.* Harrisville, NH: Total Environmental Action, 1974.

Coonley, Douglas R. *An Introduction to the Use of Wind.* Harrisville, NH: Total Environmental Action, 1975.

Coonley, Douglas R. *The Use of Man-Made Forms to Enhance the Output of Wind Collectors.* ISES Solar Energy Conference. Winnipeg, Canada, August 1976.

Eccli, Eugene, ed. *Low-Cost, Energy-Efficient Shelter for the Owner and Builder.* Emmaus, PA: Rodale Press, Inc., 1976.

Eldridge, Frank. *Wind Machines.* McLean, VA: The Mitre Corporation, 1975.

Eldridge, Frank, ed. *Proceedings of the Second Workshop on Wind Energy Conversion Systems.* McLean, VA: The Mitre Corporation, 1976.

Fales, E.N. "Windmills." In Baumeister, Theodore, and Lionel Marks, eds., *Standard Handbook for Mechanical Engineers,* 7th ed. New York: McGraw-Hill Book Company, 1967. (8th ed, 1977)

Geiger, Rudolf. *The Climate near the Ground.* 4th ed. Cambridge: Harvard University Press, 1965.

Givoni, B. Man. *Climate and Architecture.* New York: Praeger Publishers, Inc., 1969.

Glauert, H. *The Elements of Aerofoil and Airscrew Theory.* New York: Cambridge University Press, 1959.

Hackleman, Michael A. *The Homebuilt, Wind-Generated Electricity Handbook.* Saugus, CA: Earthmind, 1975.

Heronemus, William E. *Power from Offshore Winds.* Proceedings of the 8th Annual Marine Technology Society Conference, Washington, DC, 1972. Marine Tech. Soc.

Leckie, Jim, et al, eds. *Other Homes and Garbage: Designs for Self-Sufficient Living.* San Francisco: Sierra Club Books, 1975.

Lilley, G.M., and W.J. Rainbird. *A Preliminary Report on the Design and Performance of Ducted Windmills.* Technical Report C/T 119, British Electrical and Allied Industries Research Association, 1957.

National Aeronautics and Space Administration. *Proceedings of the NSF/NASA Conference on Wind-Driven Generators.* Washington, DC: NSF/NASA, 1973.

Olgyay, Victor V. *Design with Climate.* Princeton, NJ: Princeton University Press, 1970.

Park, Jack. *Simplified Wind Power Systems for Experimenters.* Sylmar, CA: Helion, 1975.

Putnam, Palmer C. *Power from the Wind.* New York: Van Nostrand Reinhold Company, 1948.

Reynolds, John. *Windmills and Watermills.* New York: Praeger Publishers, Inc., 1975.

Savino, Joseph M., ed. *Wind Energy Conversion Systems.* Washington, DC: NSF/NASA, 1973.

Technology Application Center. *Wind Energy Utilization: A Bibliography with Abstracts.* Albuquerque, NM: University of New Mexico Press, 1975.

Thomas, Percy H. *Fitting Wind Power to the Utility Network.* Report No. 099558, Federal Power Commission, 1954.

United Nations. *Wind Power.* Vol. 7, New Sources of Energy (Proceedings of the Conference at Rome, Italy, August 1961), New York: United Nations, 1964.

US Environmental Data Service. *Weather Atlas of the United States.* Detroit: Gale Research Co., 1975.

Wax, M.P. *An Experimental Study of Wind Structure with Reference to the Design and Operation of Wind Driven Generators.* Publication No. C/T 114, British Electrical and Allied Research Association, 1956.

World Meteorological Organization. *Sites for Wind-Power Installations.* Technical Note Ser. New York: Unipub, Inc., 1964.

World Meteorological Organization. *Windbreaks and Shelterbelts.* Technical Note Ser. New York: Unipub, Inc., 1964.